Quick Response
in the Supply Chain

Springer
Berlin
Heidelberg
New York
Barcelona
Hong Kong
London
Milan
Paris
Singapore
Tokyo

Eleni Hadjiconstantinou (Ed.)

Quick Response in the Supply Chain

With 57 Figures
and 7 Tables

 Springer

Dr. Eleni Hadjiconstantinou
Imperial College
Management School
53 Prince's Gate
Exhibition Rd
GB-London SW7 2PG, Great Britain

ISBN 3-540-65045-8 Springer-Verlag Berlin Heidelberg New York

Library of Congress Cataloging-in-Publication Data
Die Deutsche Bibliothek-CIP Einheitsaufnahme
Quick response in the supply chain / Eleni Hadjiconstaninou (ed.).-
Berlin ; Heidelberg ; New York ; Barcelona ; Hong Kong ; London ;
Milan ; Paris ; Singapore ; Tokyo : Springer, 1998
ISBN 3-540-65045-8

Hardcover design: Erich Kirchner, Heidelberg
SPIN 10694893   42/2202-543210 - Printed on acid-free paper

Preface

The supply chain is defined as the entire chain of raw materials supply, manufacture, assembly and distribution to the end customers. The end customers are defined here as the last point in the chain at which a product differentiation decision is made. They are the decision makers whose decisions determine an industry's success or failure; they place the orders that trigger a chain reaction. The objective of managing the supply chain is to synchronize the requirements of the customers with the flow of materials from suppliers so as to achieve a balance between the often conflicting goals of high customer service, low inventory investment and low unit cost. Supply chain management is complex involving a whole series of interlocking factors, each requiring careful consideration before cost effective integration can be achieved.

In today's competitive environment organizations are increasingly becoming aware of the need to adopt new strategies to respond to rapidly changing business conditions. Greater market share and customer loyalty can be gained by quick and reliable response to customers' changing needs. Speedy delivery of goods and services is emerging in the end of 1990s as a critical element in winning customers, retaining existing customers, and enjoying a position as an industry leader. The philosophy behind a time-based competition demands that the focus of management attention must shift from cost to time. Quick Response in the Supply Chain is the result of recent research in the US textile and clothing industry investigating ways in which the industry could improve its long-term competitiveness. It is a new corporate philosophy and a partnership strategy in which suppliers, manufacturers and distributors work together to respond more quickly to consumer needs by sharing information in order to jointly forecast future demand and continually monitor trends to detect new opportunities. Trading partners are forming closer alliances through mutual inspection of the supply chain and identifying areas of improvement such as cost reduction, service performance and time compression. From the point where decisions are taken on the sourcing and procurement of materials and components through the manufacturing and assembly process to the final distribution and after-sales support, there are many complex activities that must be managed effectively and efficiently if rapid response is to be achieved. Striving for synergy means managing a supply chain as a whole rather than its constituent elements. A whole series of trade-offs exists within individual components of the supply chain as well as across the entire range of the logistics activity. Only by understanding these interactions is it possible to develop effective strategies for managing the supply chain.

It is this strategic dimension of supply chain management based on a Quick Response philosophy that provides the focal point for the book and I have sought to select chapters that reflect the key business processes needed to achieve integrated supply chains. The material covered in the book reflects the increasing role of logistics in a global competitive strategy and explores the new concept of *Quick Response in the Supply Chain* as a way of competing, operating and management thinking. Using a number of case studies from a variety of industry sectors, I hope that the book will achieve three fundamental objectives: (1) examine the various components of the supply chain in a comprehensive manner; (2) describe ways and means of applying logistics principles and practices to achieve competitive advantage through a Quick Response strategy; (3) analyse the implications in particular areas of the supply chain giving special attention to the trade-offs that exist.

Within the above context, I have selected just six issues which I believe will be of increasing interest to senior executives beyond the year 2000:

- The Challenge of New Strategies in Integrated Supply Chains
- Systems Integration in the Supply Chain: The Enabling Technologies
- Organizational Integration: Value Adding Partnerships
- Throughput Management: Trends and Developments
- Quick Response in Outbound Logistics
- Quick Response in the International Context: Implementational Aspects.

Not all of the sixteen chapters that I have included in this book fit neatly into one of the six areas above, but between them they highlight the shift in the strategic analysis of the supply chain necessary to achieve competitive advantage and provide practical guidelines for managerial action.

Quick Response is a strategic business opportunity to alter existing business practice. Companies which reduce the time required across all business processes are reaping significant benefits and emerging as leaders in their market sectors. Much of Toyota's competitive success is directly attributable to the fast-cycle capability built into its product development, ordering, scheduling and production processes. The Baldrige award-winning companies in USA, namely Motorola, Rank Xerox and Milliken, also focus on Time-Based Management. Milliken pioneered the Quick Response Program to save the US apparel industry. Companies in many industries are operating in much the same way today. These companies make decisions faster, develop new products earlier and convert customer orders into deliveries sooner than competitors. As a result, they provide unique value in the markets they serve and they experience faster growth, higher quality products, lower costs, offer a broader variety of products and services and enjoy the best customers.

- **Faster Growth**. A Quick Response strategy enables both retailers and manufacturers to respond more quickly to changing consumer tastes. Customer service is improved through speedier processing of customer orders and reductions in overall lead times for finished products.
- **Higher quality/lower costs**. A Quick Response strategy requires that processes be simplified implying a higher quality of product or service produced at a lower cost. Everything is done right the first time. Improvements

in vendor partnerships can reduce raw material costs. There are also reductions in inventory levels and associated costs.

- The **best customers** are willing to pay a premium for a speedy delivery.
- **Greater variety**. By compressing the time required to develop and manufacture products, a company can focus on increasing the variety of products and services it offers.
- **Innovation** is a main characteristic because rapid new-product development cycles keep the company in close touch with customers and their needs.

For the top performing companies, Quick Response plays an important role in that it provides an organizational capability. The basic idea is to design an infrastructure that performs without the bottlenecks, delays, errors and inventories most companies live with. The faster information, decisions and materials can flow through a large organization, the faster it can respond to customer orders or adjust to changes in market demand and competitive conditions. Developing a strategy of Quick Response requires both an evolutionary and revolutionary approach. The materials and information flows need a degree of horizontal integration in the supply chain. This requires:

- Management of data capture and flow across the functional boundaries without delay and distortion.
- Linking systems for purchasing, production and inventory control, distribution, customer order entry and service.
- Shared ownership of information and a high degree of visibility across all functions.

The chapters in Part 1 of the book reflect this crucial contribution that a Quick Response strategy can make to competitive advantage and they propose a structured approach for the development of an appropriate logistics configuration.

Information systems are a key factor in achieving integration. Enabling Technologies are the levers which will make possible significant advances in Quick Response. Organizations with effective Quick Response strategies have exploited today's technological advances – Electronic Data Interchange (EDI), Local Area Networks (LANs), electronic mail, Electronic Point of Sale (EPoS) tills, the hand held bar code scanner and others – in such a way that the time-consuming process of communication has been reduced to a bare minimum. These facilitating computing and communications technologies enable quick, accurate, reliable and international transfer of data. Quick Response is not about technology; it is about enhancing logistical performance through the use of this technology. The use of EDI is firmly established across a huge section of commerce and industry. Retail has been one of the strongest and most effective exponents of EDI and Quick Response in the UK (e.g. Tesco, probably Europe's leading exponent of EDI). In the electronics industry, Sony's implementation of EDI has been focused on supporting their Just-In-Time (JIT) strategy. ICL as a key player in an increasingly global market has also implemented international trading relationships via EDI in many countries. In the apparel sector the integration of information systems has provided buyers with more timely and accurate sales details about colour, style and line. This too has led to more accurate forecasting, improved allocation and/or better level of sales. Integration of information ensures clearer ordering and stock control and speeds and smoothes the logistical flows.

The major theme of the chapters in Part 2 of the book is the impact of Information Technology and in particular EDI on traditional supply chain relationships and the efficiencies achieved.

The implementation of a Quick Response strategy cannot rely on a review of processes alone, but must consider systems, organization and culture. It involves commitment, enterprise-wide cooperation and long term vision. It is difficult to achieve a closely integrated, customer focused, materials flow within the traditional business organization based upon strict functional divisions and hierarchies. Instead, the Quick Response model requires broad-based 'integrators' to manage systems and people that deliver service. The focus on time broadens the scope for improvement across functional boundaries and allows organizational structures to evolve with changing market conditions. The entire supply chain is included in the improvement process. To arrive at a fully integrated supply chain, it is necessary to identify all the functions involved and to understand how they interact. Synergic companies in a supply chain are already evident in various industries, including the drugs, the automotive, the apparel and food industries. They have been referred to generally as Value Adding Partnerships and the realized benefits include improved customer service, increased market share and improved profitability. For example, the key to the success of contracting out distribution operations in view of reducing cost and improving service, particularly in those situations where dis- tribution is both complicated and critical to the client's competitive edge, is the formation of integrated logistics partnerships. The essential ingredients to achieve this are common aim, trust and information. Another example refers to the buyer-supplier relationships. The old adversorial buyer/supplier relationship can no longer be justified. A retailing concept from the USA, Efficient Consumer Response (ECR), relies on partnerships being formed between manufacturers and retailers working closely together to optimize stock levels. The level of stock reduction will depend on the confidence retailers have in manufacturers meeting their demanding service levels in terms of responsiveness, accuracy and reliability. However, it is clear that the nature of the relationship between retailers and manufacturers varies widely depending on industry concentration and the balance of power. The most significant impact that the buyer can have upon the supplier's logistics costs is through the integration of planning systems to provide the supplier with improved visibility of the buyer's materials requirements and through schedule stability to enable suppliers to optimize their own production schedules and hence minimize their inventory and working capital investment. By combining the latest developments in OR decision models, analytic databases and powerful desktop computers it is possible to construct integrated decision support systems that can be used in effective supply chain planning under uncertainty.

The chapters in Part 3 explore these issues of partnerships across the supply chain in greater detail.

One of the major implications of Quick Response is the need to assess the latest trends and developments in manufacturing logistics. Throughput management is the process whereby manufacturing and procurement lead times are linked to the needs of the market place. At the same time, its main objective is to meet the competitive challenge of increasing the speed of response to those market needs. This requires managing the supply chain as a pipeline and seeking to reduce the pipeline length or speed up the flow through that pipeline. The length of the pipeline should be only as long as the sum of the value adding steps in the supply chain process. Throughput management is therefore concerned to remove or at least reduce the blockages that occur in the pipeline. These blockages include time to build production schedules, batch sizes, machine set-up and change-over times, work-in-process, bottlenecks, excessive inventory, sequential order processing, late supplier deliveries. The major factors that affect rapid response are concurrent engineering, organization and layout, automated storage and retrieval, cell manufacturing, JIT strategies, Kanban, cross docking, warehouse management, distribution and vehicle management, demand planning and Kaizen.

The three chapters of Part 4 expand on these issues. The two case studies included refer to examples of two companies in the tyre and toiletries industries that have gained competitive advantage through their management of total throughput times and successfully applying elements of Quick Response manufacturing.

The outbound logistics process is the last interface with the customer and is important in developing profitable trading relationships. Typically, distribution can be from one or a number of geographically separate production outlets servicing directly customers or any combination of factory warehouses and national or regional distribution centres. This multi-echelon structure needs to dynamically respond to market needs through a complex communication structure that links every process in the chain. The design of distribution networks and distribution strategies is directly concerned with balancing customer service requirements (lead time and delivery performance) with issues relating to capacity and costs (inventory, warehousing, transportation). For example, reducing the number of warehouses in a distribution structure has a major effect on total distribution costs and customer service levels. Trade-offs in the supply chain are complex. There are several analytical tools available that can provide a full understanding of the implications of proposed changes in various parts of the supply chain and valuable information about where time is wasted. The type of carrier used for outbound distribution is often a source of scheduling delays. Large vehicle fleets often have to use sophisticated technology (such as Geographic Information Systems and vehicle tracking) and communication systems to improve the speed and level of responsiveness of the delivery service. A classic example of a company that has gained competitive advantage through its fully integrated time-controlled cost-effective delivery service is UPS, the large nationwide parcel operators in USA.

The case studies in Part 5 have been selected to address the issues associated with outbound logistics.

The final strategic issue discussed in Part 6 of the book refers to the implementation of Quick Response to different retail markets of the world (UK, USA, continental Europe and Japan). Success in reducing inventory through the supply chain and in minimizing lead times varies between the international markets, mainly due to the nature of retailer-supplier relations, the degree of fragmentation or concentration of retail markets, the extent of retail branding and the distribution 'culture' evident in different parts of the world.

As in Darwin's theory of evolution, those organizations that adapt to the changing world through continuous improvement ensure their survival. The overall aim of this book is to provide an insight from a variety of angles into the contribution that *Quick Response* in the supply chain can make to the achievement of competitive advantage. Quick Response is most effective as a company-wide programme in exploring new ways to gain and maintain markets in a highly competitive environment in which there is increased corporate pressure to respond to customer demands faster.

Eleni Hadjiconstantiou
January 1998

Contents

1 Quick Response: From Evolution to Revolution – New Strategies for Business Logistics

M. Walker
Logistics Consulting Services
P-E International

1 INTRODUCTION

The term Quick Response (QR) describes a number of recent developments in business logistics. QR has both specific and general logistics applications and furthermore has implications at both operations level and for company strategy.

The term QR was first used in the United States fashion apparel sector. Here information technology systems such as Electronic Point of Sale shop tills (EPoS), Bar coding and Electronic Data Interchange were combined to improve stock control and service, reduce operational costs and thereby increase profits. For companies like Wal-Mart and Target in the United States Quick Response is fulfilling the logistics dream of improved service and increased sales with a reduced stock holding. As Liam Strong, the Chief Executive of Sears plc in the UK and an exponent of QR, puts it 'we have already proved we can sell more by stocking less'. They are thereby changing the rule book which had previously been that improvement in service and reductions in cost were not compatible. They are creating a logistics revolution.

A strategy of Quick Response not only enables companies to evolve into logistically more efficient enterprises but it also changes the way in which they do business. For example, the faster and more accurate information created by Quick Response systems is enabling companies to become much more responsive to fluctuating demand. Previously these companies had bought the products speculatively and tried everything to sell them. As Strong continues 'It is essential to buy what we sell, not sell what we buy'. So QR employs evolutionary systems to create a business revolution.

More generally however the term QR can be applied to logistical improvements in a wide range of sectors. There are specific examples in grocery retailing, the apparel sector, the automotive and white goods sectors. These improvements are initiated by investments in inter, intra and indeed international company information flows.

The first key feature of QR is the faster and more accurate transfer of information both within and between all the players in the logistics chain. In all cases from fashion houses through to food retailers the impact of the business has been immense. Recent studies in the US have suggested that not only will the majority of retailers be QR retailers within a few years but that together they will achieve annual operational savings of US$ 9.6 billion by implementing these strategies. In the UK, one company, has reduced their stock holding by US$ 34 million through Quick Response.

So QR is evolutionary in the extent that it takes time to introduce systems and change. But QR is revolutionary in the scope of change and its influence upon structures, processes, attitudes and costs.

2   WHY QUICK RESPONSE

Today Quick Response is used to describe a wide variety of logistics operations. The reasons why QR has become important are many. They include the increased awareness of the costs of logistics as a key ingredient in a business operation, the importance of cycle time or service in business, the increased competitive pressures that encourage differentiation through timeliness and availability, the desire to maximize the benefits offered by new technologies and the strategic opportunity offered by all these change ingredients. This paper examines briefly all these concepts. It explores the importance of time based competition and looks at the rise of EPoS and EDI as enabling technologies. I have also attempted to fit these developments into a strategic framework.

Very few companies fully understand their true logistics or total supply chain costs. First of all most of them are solely concerned with costs over which they can have a direct and immediate influence and secondly few can define or account for their logistics sufficiently clearly to be able to allocate costs.

Furthermore physical distribution costs which are their main concern are very much a function of the product and the service offered. For example a supplier of packaged groceries to a food retailer might spend 2.5% of supplier sales revenue on warehousing, a further 2% on transport, an additional 1.5% on administration and order processing and the equivalent of 1% will have been spent on stock financing. So in total 7% of the delivered cost of the goods has been incurred by the supplier. The retailer will then impose a margin and re-distribute it to their shops. A retailer will incur additional distribution costs of perhaps 5% of sales. If we weight all these costs to eliminate the differences caused by changes in the retail price or margin arrangements it is easy for the final supply chain cost to be 10% of the product's consumer price. Yet in sectors with higher service require-ments such as car parts distribution costs can be up to 25% of sales. So if we define logistics as only being appropriate to finished goods distribution we can see that it is of considerable importance and worthy of management attention.

Traditional distribution cost studies have tended to concentrate on incremental improvements to operational aspects such as transport and warehousing. Over the course of time it becomes increasingly difficult to produce substantial further savings in these areas. With mature distribution networks and a free market in transport, companies will be looking to other areas of cost for savings. Consequently the impetus will be concentrated in two areas.

Firstly to examine the areas of inventory reduction and more cost effective order processing policies and costs and secondly to explore efficiencies in the whole logistics chain rather than just internal distribution chains. This strategic shift in analysis makes Quick Response with its emphasis on speed, inventory reduction, information chains and partnerships the focus of attention and the route to future success.

Understanding and addressing the key costs of logistics is one factor of concern to management but so is the need to consider time. There is increased corporate pressure to respond to customer demands faster. Yet many companies underestimate the time taken to process an order internally. A recent study by my company for a leading European manufacturer revealed that the time involved with order capture and processing was five times longer than the finished goods transit time. So management emphasis placed on improving both communications between companies and internal processing time would help meet increased customer requirements.

Today QR is taking place rather than just being talked about. In the 1970s and 1980s many companies were slow to invest in information technology. The reasons were readily apparent, poor hardware performance, the fear of being locked into proprietary products, the problems of change and upgrading resources, the lack of compatibility with other users and an inflexible approach to growth and development on the part of their platform providers. Much of this has now changed as a result of improvements in information technology. With the availability of cheap high performance computing power, the evolution of the PC into the Electronic Point of Sale (EPoS) till, the hand held bar code scanner and the integrating revolution brought about through 'open systems' and Electronic Data Interchange the elements necessary for the Quick Response revolution has begun.

Quick Response relies upon the building of an information chain between the manufacturers mill, though distributors and suppliers to the retailers till. Elements of this information chain will include Universal Product Coding (UPC), Electronic Point of Sale (EPoS) shop tills, handheld scanners or bar-code readers, Electronic Data Interchange (EDI) networks, forecasting software, agreements on data exchange protocols, the use of Value Added Networks (VANS) and increased information sharing. But QR is not about technology it is about enhancing logistical performance through the use of this technology. It requires vision and a willingness to change based upon the creation of a seamless and transparent information chain. And that is a critical element of the revolution.

3   ELECTRONIC POINT OF SALE

Typically this information chain begins at store level because another element of the Quick Response revolution is about creating a demand driven logistics chain. Here many retailers are investing heavily in Electronic Point of Sale equipment.

The rate of EPoS installation in recent years has speeded up. Many companies are now engaged in a four or five year rolling programme to invest in EPoS equipment. In the mid 1980s the Article Numbering Association, reported that Boots the Chemist had two stores fitted out with EPoS equipment. These were essentially test bed operations. By the middle of 1992 all their stores were fitted with EPoS and they had invested in over twelve thousand tills. In the process they have invested over £70 million with an annual operating cost of £8 million. Yet their director for information services states that in 1993 these systems will account for £90 million worth of profits.

Other companies particularly those in the grocery sector have witnessed extensive growth in EPoS. Companies such as Safeway and J Sainsbury have now installed EPoS in virtually all their stores. Tesco, for example, have scanning in 82% of their stores and they account for 94% of their sales. Indeed in the country as a whole 70% of grocery sales are now sold across scanners having grown from only 2% in 1984.

4 ELECTRONIC DATA INTERCHANGE

The increase in the installation of EPoS tills is matched by growth in the number of companies using Electronic Data Interchange (EDI). Here data is moved between one company's computer application and another in an agreed format and usually via a third-party. Companies like British Telecom, IBM, Istel and INS are all supplying inter-company links or Value Added Networks. By 1995 it is estimated that the UK EDI user base will exceed 15,000 companies. Again the grocery retailers are substantial users. Tesco recently announced that they already have 1,300 suppliers on their EDI system. For some EDI has already become the normal way of life. B&Q, a leading UK home improvement retailer, have imposed EDI as a condition of trading on its suppliers. Toys R US, the US retailer with a rapid expansion programme in Europe, has a policy of charging suppliers US$ 25 every time it has to process a piece of paperwork.

The benefits of EDI from a recent survey ranked faster transactions, reduced administration and improved service as the three most important features. Reductions in stock levels and partnership development came later. Clearly EDI is currently being used to improve existing processes by speeding them up. Food retailers can now place orders and receive deliveries within 16 hours. The high street pharmaceutical dispensaries can process orders and receive them in a few hours. This is only possible with EDI. But saving time is not the only benefit. Cost savings too can be substantial. Sears in the US demonstrated that it costs them US$ 52 to process a paper order and US$ 5 to process an EDI order. One UK company was able to reduce their telephone charges by nearly US$ 2 million as a result of installing an EDI system.

Yet whilst the growth in numbers of companies using EDI is substantial there is even greater potential in increasing the uses of EDI. Those companies that are using EDI are failing to use it to its full potential. EDI can be used to exchange a wide variety of information necessary to improve enterprise logistics. This information can involve orders, stock availability, product prices, stock availability, forecasts and to communicate via e-mail.

However, the full logistical system benefits of reduced stock levels are not yet widely enjoyed. Yet they are available. A comparative study of two blue chip UK based manufacturers, both of whom sell similar products to the same customers, demonstrates the clear benefits. One company is an extensive user of EDI and renowned for its use of information technology. The other relies heavily on the telephone and indeed the postal service. The non-EDI user spends twice as much on order processing as the EDI user. The EDI user can control stock levels in a better way so that again the rival ends up holding twice as much stock as the EDI user. In all these differences accounted for a cost difference of nearly US$ 8 million per annum on distribution cost.

5   THE ROLE OF THE STRATEGIC ALLIANCE IN QR

Strategic alliances are typified in the UK by agreements between the grocery manufacturers and retailers such as Tesco. And they do bring significant results. As the jointly produced forecasts get better so safety stock cover and ultimately cost levels can be reduced. In a recent example this partnership led to total stock being reduced on one occasion by 50,000 cases and produced a cost saving of nearly US$ 600,000. Between them the two companies have reduced the average stock holding on certain lines from 1.79 weeks to 1.04 creating a truly lean logistical operation between two players.

In the apparel sector the integration of the information chain has provided buyers with more timely and accurate sales details about colour, style and line. This too has led to more accurate forecasting, improved allocation and a better level of sales. The extent of markdowns, the garments that have to be discounted after their selling season, has been reduced by up to a half in some cases. And for these same companies stock availability has also improved from 75% to 95% and beyond. So they are selling more by stocking less. This is a truly revolutionary approach.

So Quick Response builds upon the technology and information supplied by the rapidly expanding EPoS and EDI systems. But QR is more than just operational issues. Quick Response starts at the adoption of automation in the supply chain process and continues through to manufacturing companies. For companies like the jean manufacturer Levi-Strauss and General Electric, Quick Response has become their corporate philosophy.

6   CONCLUSION

QR can be applied in most sectors and to most products. However, technology aside the issues are ones of suitability of product and the ability to form partnerships with the other players. QR is about forming an integrated information chain for 'till to mill'. It involves integrating retailer information chains with distributors and suppliers. Integration of information ensures clearer ordering and stock control and speeds and smooths the logistical flows. The rewards of QR will be less stock holding, faster ordering and delivery, a reduction in costs and increased profits. With this timely and more accurate information some considerable benefits are available in terms of more accurate stock control and forecasting. Indeed we will rapidly see the introduction of advance shipping notices as warehouses are advised of inbound movements combined with product tracking systems and automatic replenishment systems. Other improvements to current practice will also come about. These might include EDI invoicing, self billing procedures, joint forecasting or even allowing the manufacturer to undertake the re-buy function. Simply put Quick Response involves changing the way a company undertakes its business. It requires a shift from the traditional adversarial roles to a more open partnership arrangement. And the bond in this partnership is the information link.

REFERENCES

Baker, R. (1991) *The Long Path to Quick Response EDI*, McGraw-Hill.

Chain Store Age Executive (1991) Quick Response, The Path to Better Customer Service, 3.

Christopher, M. G. (1971) *Total Distribution*, Gower Press.

Davis, H. (1992) Quick Response Now, *Davis Database*, 3.

Dolen, P., Grottke, R. and Lucker, J. (1990) *Quick Response: A Cost Benefit Analysis*, Arthur Andersen.

Electronic Trader (1992) November/December.

European Logistics Consultants Distribution Cost Database (1992).

Hopwood and Carter (1988) EPoS and its Impact on the Retail Environment, in: (Ed. A. West) *Handbook of Retailing*, Gower.

Institute of Logistics Annual Distribution Cost Survey (1992).

Kramer, S. (1991) *Quick Response, An Implementation Guide*.

Metzgen, F. (1990) *Killing the Paper Dragon*, Heinemann.

Morton, S. The Corporation of the 1990s.

PFA Research Ltd. (1992) UK State of the Nation, *Electronic Trader*, 10.

Stalk, Jr G. (1991) The New Manufacturing, Time the next source of Competitive Advantage, *Harvard Business Review*.

Supply Chain Partnerships Institute of Grocery Distribution Annual Conference 1991.

Walker, M. (1992) The Distribution Implications of Quick Response, Quick Response Conference, London.

2 The Quick Response Model and its Applicability in the UK

M. Silman
IBM Global Services, Network Services

1 INTRODUCTION

Quick Response developed in the US textile and clothing industry as a result of research, instigated in 1984, by the 'Crafted with pride in USA council'. The research was intended to investigate ways in which the industry could improve its long term competitiveness and focussed on supply chain analysis. It revealed that although individual components of the supply chain were efficient, the overall efficiency of the system was very low. In seeking to minimize costs independently of each other the fibre, textile, clothing and retail companies were inadvertently pursuing practices that added significant costs to the overall supply chain.

The clothing supply chain, from raw material to consumer purchase, was 66 weeks. Of this, 11 weeks was in-plant time, 40 weeks was in warehouses or transit and 15 weeks was in store. This long supply chain was both expensive to finance and, even more significantly, resulted in major losses as either too much or too little product was produced and distributed based on inaccurate forecasts of future demand.

The research led to the development of the Quick Response (QR) strategy for merchandise retailers and suppliers. QR is a partnership strategy in which retailer and supplier work together to respond more quickly to consumer needs by sharing information on Electronic Point of Sale (EPoS) activity to jointly forecast future demand for replenishable items and to continually monitor trends to detect new opportunities for new items.

The current usage of QR can be represented by a 5 stage model representing increasing levels of sophistication and partnership. As companies progress towards later stages, they strive to attract and construct mutually beneficial alliances and partnerships for the long term.

2 STAGE 1

The initial stage tends to concentrate upon the technical infrastructure required and as such, QR tends to be regarded as a technology based initiative rather than as an organizational one.

Typically, stage 1 includes the implementation of such technologies as:

- Article Numbering/Universal Product Coding (UPC)
- Point of Sale (POS)
- Electronic Data Interchange (EDI)
- Shipping Container Marking (SCM)
- Online UPC catalogue.

This list is not exhaustive and it is not meant to imply that without all of these technologies, progression to later stages cannot be achieved.

3   STAGE 2

The second level of sophistication relates to the automatic replenishment of sold goods. Here, the retailer and supplier commit themselves to having known lead times for replenishment which are shortened as much as possible.

A further development is that the retailer can use automatic replenishment techniques where purchase orders are automatically generated and forwarded to suppliers. Known lead times are fed to the system which can calculate when to place an order considering on-hand inventory levels.

Many UK companies will claim to be operating in this environment already as they can supply to retailers in 3–4 days typically. However, they can only achieve these service levels by supplying from stock; which inevitably increases costs and decreases margins.

4   STAGE 3

The third stage is concerned with trading partners forging closer alliances through mutual inspection of the supply chain and identifying areas of improvement. The scope of inspection is limited to the replenishment of existing products and concerns areas such as cost-reduction, service improvements and time compression.

One of the more common examples of stage 3 operation is forecasting, where mathematical algorithms are applied to historical sales data to determine future demand.

In the traditional operating environment, forecasting is done by both trading partners independently of each other using different forecasting models, which results in different order requirements being predicted causing confusion and increased costs in the chain.

In the QR environment, both partners use the same data in conjunction with each other to anticipate future demand. This sharing of data is a radical departure from current retailer-supplier relationships where an adversarial environment is the inevitable consequence.

5 STAGE 4

QR takes it name from the goal of responding more effectively and more quickly to consumers changing tastes. As the market becomes more fragmented and more volatile, the traditional sequential product development process becomes less and less effective.

This traditional process is organized functionally, with each function doing its own task before handing over to the next. Whilst this may be beneficial in terms of project management, it also creates a long and inflexible process with excess product marked down to be sold at or below cost and so reducing margins.

QR companies have made major changes to their product development processes. Their new method is based on continuously monitoring PoS activity to detect trends and constantly testing new product concepts in a limited number of stores. Due to the importance of speed, these companies have created cross-functional development teams (containing members from all companies involved), containing all the functional skills required, e.g. design development, costing and purchasing.

Working concurrently on a new concept the team members can cut weeks out of the development cycle.

6 STAGE 5

The fifth stage of the QR process model relates to a partnership scenario where the supplier is handed full responsibility of the supply chain right up to point of purchase by the customer.

Here the supplier is given an allocation of shelf space by the retailer and the supplier is charged with providing full stock replenishment, in-store promotion and staff product-training where necessary.

This is not necessarily the final stage of the QR process. It is likely that leading QR companies will be adding further stages to the model when they have fully understood and appreciated the benefits of their existing QR based processes.

7 UK MARKETPLACE

Unlike the US marketplace, the UK is dominated by 5 or 6 large retailers, predominantly based in the grocery sector. Whereas in the US retail margins tend to be of the order of 1% and cost pressures are high, the UK supermarket environment operates typically with margins of 6–7%.

Additionally, 'own brand' products constitute a significant portion of the overall business in this sector, whereas in the US, this concept is barely used.

The result of these differences is that UK retailers are far more powerful than their US counterparts and they tend to adopt a dominant role in relation to their suppliers. In the short term, this is a disincentive to the adoption of QR, as the existing organizational adversity – retailer versus supplier, can be continued in the adoption of EDI and UPC, etc. purely for the benefit of the retailer and often at the expense of the supplier.

However, outside of the grocery sector, power tends to reside more with the suppliers. Therefore, QR will be more readily accepted.

Longer term, with the introduction of American style warehouse clubs the cost pressure driving US industry will become evident in the UK and the flexibility and profitability achieved in other sectors as a result of QR practices will be identified and transferred to the grocery environment.

Technically, the availability of real-time EDI and multimedia shopping services will further enhance the benefits achievable from QR implementation, coupled with customer loyalty schemes which will make customer targeting and product development even more sophisticated as the marketplace becomes more fragmented and volatile.

As the customer moves into a 'networked world' via the Internet, the pressure for QR to be implemented and expanded will become greater. A balance will be demanded between 'point and click' instant response shopping, and a similar response time for fulfilment. This has implications for the entire supply chain.

The logical conclusion is that the alliances formed in order to enable QR will result in redefined corporate relationships, boundaries and entities – with many of the slower players dropping out altogether.

3 The Role of Logistics and IT in the European Enterprise

G. Stevens
KPMG Management Consulting

1   INTRODUCTION

The shape of European manufacturing and distribution must change. Historically, multi-national manufacturing and distribution companies operating in Europe, particularly those involved in the supply of high-tech and fast moving consumer goods, have comprised national sub-units with performance measured by local bottom line results. Communication and interaction between sites has been constrained by import duty barriers, quotas, national standards and regulations. This is changing. The move towards a single market, changes to statutory requirements and the introduction of the concept of 'mutual recognition' has provided companies with an opportunity to standardize products and business processes across Europe. The Schengen Agreement has progressively simplified border crossing between France, Germany and the Benelux countries, reducing border crossing time and accelerating the movement of products.

The competitive environment in which European companies operate is changing. Markets are global and competition is accelerating. Global sourcing provides the flexibility companies need to steal a march on the competition. It's a seller's market. Requirements for delivery time and reliability are becoming more demanding. Pressure is on to reduce costs and increase the quality of products and services. Just-in-Time and computer based material planning systems force suppliers to supply small orders with fast, reliable response times. The impact of these changes is that the limitations inherent in a nationally focused business will severely jeopardize the performance of many of today's leading businesses. Companies are being forced to develop a multi-national perspective and are having to review the disparate elements of a European organization; restructuring them into a single, European Enterprise.

2   THE ROLE OF LOGISTICS

All over Europe, we are seeing rationalization of product ranges across sites. However there are dangers and drawbacks from manufacturing a product on only one site. The more focused the facility, the greater the potential risk to the organization, the more vital the effective running of the plant becomes and the greater the reliance on the logistics infrastructure. Reliability and quality of service are just as important as price for winning Just-In-Time business. Companies must now turn their attention to the more humdrum but equally essential issues of logistics, including purchasing, materials management, warehousing, and distribution. They must recognize that an efficient and effective logistics operation can serve as a means of achieving competitive advantage. The majority of current logistic systems are not good enough. To quote one European Logistics Director: '80% of multi-nationals in Europe are looking to change their European logistics systems'. The logistics function is having to change its role. No longer is it responsible just for shifting boxes between storage points. It must now operate as 'manager' of the complete supply chain; co-ordinating all the activities concerned with the flow of material from suppliers through the manufacturing process and distribution network to customers.

Competition is not restricted to European organizations – Japanese manu-facturing companies entering the European market are playing a significant role. Companies such as Sony and Toyota have been developing green field sites. This has allowed them to exploit the use of available logistics networks to best effect, giving them a significant competitive advantage over many indigenous producers who have developed their European facilities on an *ad hoc* basis. The Japanese have realized that effective logistics management can be used as a means of achieving competitive advantage. This is illustrated, not only by the commitment being displayed by Japanese companies, but also the number of Japanese logistics companies setting up in Europe in support of these companies. Efficient use of stock can significantly reduce costs and release the productive potential of manufacturing facilities. The amount of capital tied up in inventory in the UK is estimated to be more than £80 billion, representing about 20% of the value of manufacturing output. In Japan, the corresponding figure is 10.5%.

Essential to the operation of an effective European Enterprise is the develop-ment of an appropriate logistics configuration. Developing a suitable configuration requires a structured approach.

Define the customer service mission
First, the customer service mission of the organization must be defined. Key elements of the customer service offer should include: lead time, delivery frequency, reliability and first time order completeness. These will drive the logistics structure. Actual customer requirements must be established. Trying to second guess the customer is likely to lead to disaster.

Profile the existing operation
The profile of the existing logistics infrastructure should be determined. Data on plant and warehouse locations and characteristics needs to be collected. The historic and likely future goods flow through the demand chain needs to be con-sidered. Existing logistics costs must be identified and related to the activities in the organization. The application of techniques such as activity based costing are particularly relevant to this task. This is often the first time that the organization

has separated its logistics cost on a pan-European basis. It is not uncommon for these costs to be in excess of 10% of total turnover.

Consider alternative configurations
The various alternative configurations need to be considered. The wide range of options available should be brainstormed before narrowing them down to the preferred approach. The use of a simulation model to assess the implications of each option is a particularly powerful tool. The model can be used to calibrate the existing configuration as a basis for benchmarking the alternatives. It should be possible to model a number of 'green field' options to establish the ideal configuration. Any additional costs incurred by the use of existing facilities can be overlaid to allow the company to arrive at a balanced decision.

Explore the use of third party contractors
Before detailed design and implementation is undertaken the tactical decision as to which elements of the logistics infrastructure can best be supported by third party contractors need to be taken. Suppliers of logistic services are beginning to operate on a pan-European basis. Many are looking to form partnerships, taking on the role of 'front man' for complex transport and distribution schemes, encompassing responsibility for management of both inward and outward logistics.

3   THE USE OF INFORMATION TECHNOLOGY

Operation of an effective European Enterprise based on a European outlook, incorporating a wide range of inputs and end points, involves the co-ordination of multiple activities on a timely basis. No longer can an organization rely on informal communication. The physical distance between the elements of planning and control invariably preclude face to face contact. Experience has shown that an effective information technology (IT) infrastructure is key to the development of a successful European Enterprise. The new information networks need to be integrated both internally and externally to link systems together along the entire supply chain (suppliers, production sites, distribution centres and customers). A number of leading edge companies, such as Sun Microsystems, Gillette, Colgate-Palmolive and Tambrands, are already developing integrated information systems to support decision making, co-ordinate vendor to customer movement, and share information on production location and status.

Only recently have manufacturing and distribution systems started to come onto the market place able to meet the operational needs of the European Enterprise; in particular, the need to operate in a multi-plant, multi-currency, multi-language and multi-legal entity environment. The scope of a suitable system must include both manufacturing and distribution – in other words, sales order processing, resource and material planning, inventory control, purchase order processing and distribution requirements planning. Effective implementation of an integrated information system within a European Enterprise allows companies to make supply, production and distribution decisions without regard to national boundaries, whilst at the same time satisfying the requirements of different legal entities in the countries in which they operate. Different information requirements that flow from this type of situation include:

- *Forecasting*
    - collation of forecasts from different markets, countries and product groups and/or item level.

- *Logistics/inventory planning*
    - aggregating market/country stocks to derive European inventory,
    - netting demand against inventory to determine market/country and European net requirements,
    - flagging out-of-balance demand/inventory situations at both market/country and European levels,
    - providing inventory deployment 'what if' capability that identifies and supports intercompany/transfer opportunities,
    - providing shipment schedules to individual site systems,
    - allocating orders to particular supply points,
    - tracking the progress of orders.

- *Product data*
    - common product definitions,
    - central updating of product data, bills of materials, formulations, process details etc. that reside on remote systems.

- *Production planning*
    - aggregation of forecast, sales order and finished goods levels across all remote sites,
    - the ability to link demand by item or customer to preferred site of production,
    - rough cut capacity planning and 'what if' simulations across plants and key production resources,
    - splitting and down loading of aggregate production plans to remote sites.

- *Purchasing*
    - feeding of some or all purchase requisitions from remote sites to a central processing function,
    - updating of purchasing orders placed by remote sites,
    - feeding of some or all goods received data at remote sites to central processing,
    - invoice processing either centrally or at remote sites.

- *Product costing*
    - analysis of manufacturing costs across all European production facilities,

   – managing information analysis of sales, stock turns etc. across multiple
   systems.

Businesses that want to implement integrated information systems need to
address a number of key issues. Unless these are addressed explicitly they will
undermine the performance of the European Enterprise. These include:

- *What is the most appropriate level of European integration?*
   – The greater the degree of the integration, the more complex and costly
   the likely solution. Organizations which are required to manage a har-
   monized product range, using common production processes across
   Europe will need to be able to make decisions on sourcing and resource
   allocation on an almost daily basis. In this situation a high level of
   integration is essential. If the product range is disparate, with unique
   production facilities and only limited flexibility across Europe, then a
   lower level of European integration is likely to be adequate. In this
   situation sourcing and supply decisions are likely to be made as part
   of an annual or quarterly planning cycle.

- *What project structure should be put in place?*
   – It is important to establish a project team able to co-ordinate local activities in line with European requirements. This European team should
   comprise user, business and IT representatives.

- *How should the communication between sites be managed?*
   – The project team should be supported by a series of local taskforces
   responsible for ensuring that the needs of each site are identified and
   addressed. Communications between the various sites should in general
   be via the European project team.

- *How should the inevitable conflicts between local and European requirements be
   resolved?*
   – Part of the responsibility of the European project team is to ensure that
   the needs of the local facilities are taken into consideration when developing the European systems. The extent to which local requirements
   are able to be accommodated is directly related to the level of integration. An integrated system is likely to demand a higher degree of
   compromise on the part of local operations.

- *How should the development of systems required to ensure European integrity be
   controlled?*
   – As the business develops inevitably the systems will need to be
   enhanced. Unless a 'core' system is established such that it can be
   enhanced in accordance with the changing needs of the business, it will
   be almost impractical to co-ordinate a variety of local developments.
   The 'core' system should include only those elements which are
   essential to effective operation of a pan-European business. Peripheral

activities which do not impact the European perspective can invariably be satisfied at the local level.

- *What is the most appropriate set of performance measures?*
  - The key benefit arising from the operation of a European Enterprise is the ability to manage 'inventory tension' within the supply chain. There is the tension between inventory investment, service level and unit costs. The co-ordination between these conflicting forces has to be underpinned by consistent performance measures. The key measure of effectiveness is likely to be service level. Objectives should be set and performance measured at various points within the chain, including the movement of material into and out of the distribution network.

- *How should data be defined and maintained to facilitate consistent reporting?*
  - One of the biggest obstacles to product harmonization and pan-European reporting is lack of consistent data formats and definitions. It is impossible to develop a view of the European market by product group unless there is consistent definition of market and products.

- *What is the most appropriate hardware and database topology?*
  - Decisions must be made on the means of holding, processing and distributing information throughout the European Enterprise. This may involve the use of wide and local area networks, distributed databases, common hardware and software platforms and interfaces between a variety of central and local applications.

- *How should training and education be managed to ensure consistency of operating practice across sites?*
  - One of the key tasks of the European project team is to provide core training and education to all parts of the European Enterprise to ensure a common approach and operation. This needs to be supplemented by local training given by local staff.

4   INVENTORY PLANNING AND CONTROL

It is necessary to continually take decisions on the allocation, location and movement of products. A key issue is who should make these decisions? A European co-ordinating function should be established to act as a 'clearing house' able to provide a view on the impact of a decision on net profitability to the European Enterprise rather than to the individual subsidiary. The function can be passive or active. If passive, it acts merely as a clearing house for accumulating subsidiary orders and passing them onto production. It is responsible for ensuring that a subsidiary receives what it forecasts. The forecast must be received ahead of the latest manufacturing start times. The ability to respond to short term changes in demand is minimal. If active, it has a responsibility of trying to achieve the best trade off between product availability, stock levels and distribution costs. In reality although a suitable entity can be small in terms of the number of people involved to be effective, it must have teeth.

Despite improved systems leading to reduced lead time and better forecasting and demand, there will still need to be stock buffer somewhere – usually known as 'strategic' stock. This means that in the majority of cases stock allocation should not be a problem. Following the 80/20 principal there is however likely to be some products where for manufacturing or demand reasons there is insufficient stock. Stock allocation decisions consequently have to be made at a European level. This is invariably a difficult process requiring trade-offs to be made for the good of the European Enterprise. It is important that subsidiaries understand the rationale behind the rationalizing process.

5   CONCLUSION

The development of the European Enterprise presents a significant opportunity for a number of organizations. Potentially it offers a 'step change' in performance, with the opportunity to substantially reduce the levels of inventory across Europe and correspondingly increase service levels. Whilst the potential benefits are high so is the cost of failure. No longer is it an isolated national operation that is put at risk. The potential cost of failure is a loss in performance of the European business. Companies moving down this route are pioneering. The ones that succeed are those that are prepared to recognize, and address, the key issues and build on the experience of others.

CASE STUDIES

1   TAMBRANDS LIMITED

Tambrands manufactures fast moving consumer goods on three sites across Europe. It has a turnover of $185 million out of corporate sales of $670 million. Within Europe Tambrands operates across three main manufacturing plants in England, Ireland and France. Currently, these plants operate as autonomous units. Tambrands have recognized the inefficiency of the current structure and are in the process of re-organizing their European operations and are moving towards the goal of 'The European Enterprise'. Ultimately, Tambrands will run their European operation as a single entity, synchronizing distribution to retailers from linked warehouses and in turn replenishing the stocks from manufacturing facilities under the control of a single, consolidated production schedule. To quote Peter Napier, Senior Vice President – Operations.

'Tambrands is dedicated to a pan-European structure in all of our manu-facturing operations. From multi-national packs, product source from any factory to any market, European vendors negotiated on a pan-European requirements basis – common product packaging specifications, to inte-grated inventory consolidation over multiple factories, central warehousing and customer stocks. Our new system will address all of the elements of a multi-currency, multi-market, multi-manufacturing environment with an on-line multiple facility database. Only through installation and optimization of such a system can we maintain our competitive advantages'.

The objective: to select an integrated software package encompassing dis-tribution, manufacturing and financial accounts, which could be implemented throughout Europe in support of the company's pan-European business strategy (see Figure 3.1).

The process: the company's European project team generated a statement of pan-European business requirements. This required significant commitment and an effective and efficient communication structure to ensure adequate input from the individual European sites.

Selected vendors were invited to demonstrate their systems' fit with the company's requirements. During the process it became apparent that while the vendors could offer multi-plant solutions, they were unable to truly support the company's pan-European vision of both manufacturing anywhere in Europe and distributing throughout Europe. Specific European requirements that a single version of the software should support included multi-standard cost, multi-currency, multi-legal entity, pan-European inventory deployment and distribution.

Initially the software vendors' development staffs did not appear to appreciate fully the requirements of trading in a pan-European environment across several legal entities. However, by working with the vendors, the company was able to establish the necessary functionality a true pan-European manufacturing software package should possess. This base functionality was then used to specify required software enhancements necessary to support the company in the short term. More important, the base functionality was used to check that the vendor's long-term development plans would deliver the required functionality as part of their standard package.

The selection process was judged a success not simply because a vendor solu-tion was selected that is capable of supporting the company's current and future European trading vision, but also because user ownership of the system has been readily accepted across all European sites – thereby providing firm foundations for a successful implementation.

Business benefits:

- reduced inventory holding throughout the entire distribution chain from centralized planning,

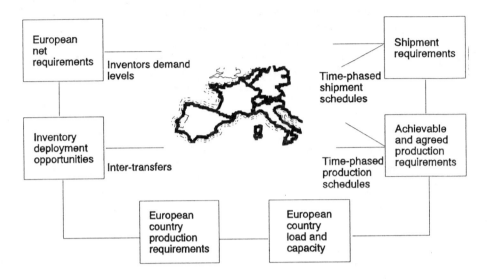

FIGURE 3.1. Supporting systems for pan-European trading.

- improved customer service resulting from pan-European inventory deployment,
- reduced manufacturing costs by enabling the lowest-cost production site to be identified and exploited,
- enhanced purchasing leverage.

System benefits:

- the ability to utilize key systems personnel on a pan-European basis,
- shared expertise, experiences and development,
- opportunity to benefit from corporate purchasing (e.g. the company negotiated a reduction on the software purchase price of more than one-third).

The company is now in the process of implementing the chosen solution.

2   SUN MICROSYSTEMS

Sun Microsystems is a leading manufacturer of UNIX workstations and systems. KPMG has recently completed a major project to assist Sun replace their European sales and distribution systems. The first part of the process was to develop a pan-

European business model, which challenged the existing business processes in Sun and where necessary introduced new business processes. The project coincided with a major change in physical distribution of product throughout Europe. Instead of holding inventory in each of the ten Sun European subsidiaries, inventory is now consolidated centrally in Holland and in many cases is now shipped directly to the end customer from there. The new sales and distribution systems strategy was based on decentralized systems throughout Europe, linked to a centralized worldwide manufacturing system.

4   Electronic Trading in Europe

R. B. Cole
Independent Consultant

1   INTRODUCTION

The intention of this chapter is to examine the potential means of delivering pan-European electronic trading systems and the prospects and pitfalls awaiting those endeavouring to trade electronically across Europe.

The primary objective of the European Economic Community is to evolve from diverse and incompatible economies a single European market allowing open and unrestricted competition in all participating states. This is creating probably the world's largest single unified market and presents the member state domestic industries with an expanding market exceeding 350 million consumers within which they must compete. It provides the possibility of industrial economies of scale previously restricted through the historically fragmented markets and nationalistic attitudes. Removal of restrictions is intensifying local market competition and forcing companies to consider how to extend into the wider market whilst protecting their local customer base. Operation within this marketplace requires redefinition of both the ground rules and business concepts with supply logistics chains tuned to quickly and cost effectively meet new challenges. Delivery of services meeting the quality and cost level criteria of customers is now key to survival for each member within a supply chain; and electronic trading is emerging as a key element in achieving the desired standards.

```
• FOR:              EC bureaucracy
                    Company administration
                    Product handling
                    Local Government

• CAN BE:  Inaccurate
                    Inappropriate
                    Late
                    Lost

• RESULT
         Synchronization
         Delays
         Safety of goods endangered
         Investment costs incurred
         Diminshed perceived service quality
         Cash flow impacted
```

FIGURE 4.1. Information flows.

```
┌────────────────────────────────────────────────┐
│                                                  │
│   • LOGISTICS CHAIN CONFIDENCE                   │
│                                                  │
│   • PARTNERSHIP TRUST                            │
│                                                  │
│   • SHARE    INFORMATION                         │
│              FLOW RESPONSIBILITY                 │
│              ADMINSTRATION                       │
│              RISK                                │
│                                                  │
│   • AIM      INFORMATION TRANSFER                │
│              RAPID                               │
│              ACCURATE                            │
│              RELIABLE                            │
│                                                  │
└────────────────────────────────────────────────┘
```

FIGURE 4.2. Trading reach requirement.

In transportation provision of rapid physical movement of goods throughout Europe to meet tight service level requirements is already evident. However the accompanying information flow, whether for EC bureaucracy, company administration, product handling or local Government, regularly inhibits or stops the transportation process. The causes may be inaccurate, inappropriate or late documentation. The consequential delays through documentation synchronization failure in a supply chain endangers the safety of goods, incurs investment costs, diminishes perceived service quality and impacts cash flow. For suppliers to believe they can provide the 'reach' necessary to trade across Europe they must be confident in all the logistical elements in the chain. Hence they must be prepared to foster close partnerships within that chain to minimize delays and risk between the conception, production and consumption of goods. These liaisons require trust and a willingness to share in the information flow, administration processes and risk.

A prerequisite to achieving the close co-operation is the provision of a means of rapid, accurate and reliable transfer of information to assist the free circulation of goods and services. The latter requirement implicitly calls for a 'European Electronic Trading' infrastructure that complements the market and efficiently integrates '*Electronic Communication*' into the business processes of the European Community. The objective is a common electronic infrastructure on which the entire spectrum of commerce feel comfortable operating according to their specific requirements and which also incorporates the flexibility for competition and innovation.

The removal of internal borders leaving a business community facing a new 'VAT-rich' environment emphasizes the need for an electronic back bone for Europe. The Commission's DG XXI uniform computerized system for VAT registration and VAT summaries demands well conceived computerized systems exploiting electronic communications to meet the anticipated tight time frames. These systems with their need for complementary Electronic Data Interchange (EDI) software are an impetus to Electronic Trading growth as the various Revenue

```
┌─────────────────────────────────────────────────┐
│                     DGXXI                         │
│                                                   │
│         VAT Information Exchange Systems           │
│             VAT EC Sales Listings                 │
│             INTRASTAT returns                     │
│                                                   │
│                     DGXIII                        │
│                                                   │
│      Trade Electronic Data Interchange Systems     │
│                    (TEDIS)                        │
│       Research, educate, encourage, cajole, guide  │
│                                                   │
│                 PHASE 1 = ISSUES                  │
│                 PHASE 2 = RESOLVE                 │
│                                                   │
│                  EEC INTIATIVES                   │
│                                                   │
│         TELEMATIQUE                               │
│         X435 CONNECTIVITY SERVICES                │
└─────────────────────────────────────────────────┘
```

FIGURE 4.3. European Commission 'Stick and Carrot' approach to ET.

agencies publicize their requirements for the VAT EC Sales Listings and INTRASTAT returns.

The European Commission has long recognized the importance of electronic data communications to Europe and through DGXIII focused on the issues in providing a common infrastructure. Progress has been slow despite a number of projects and the initiatives of their TEDIS programme (Trade Electronic Data Interchange Systems). The continuation of the TEDIS programme into a second phase with additional capitalization suggests the Community approach to Electronic trading is to be maintained, although the extent and depth of sustained effort needed to educate, encourage, cajole and guide participants appears to grow rather than diminish. The major result of phase one of the TEDIS programme was the identification of a range of issues pending resolution, phase two must now resolve them. Certainly, if businesses in the peripheral regions of Europe are to survive they need to be able to offer cost effective services to their trading partners throughout the business life cycles, and it is difficult to see this being achieved without electronic communications and encouragement and support from initiatives such as Telematique (see Figure 4.3).

To comprehend the enormity of the challenge in creating an open and effective electronic marketplace it is necessary to understand the elements that constitute Electronic Trading and control its growth. A second phase is acceptance that achieving the goal necessitates individuals and communities abandoning cherished systems, localized standards and previous investments in skills, time and money. Specialists who have encountered the 'not invented here' syndrome know the strength of resistance to adopting new standards and processes, especially when there are close personal involvement in superseded systems. To overcome emotive issues the arguments for new procedures need to be overwhelming. For many companies and individuals this will probably be unachievable until electronic trading within international markets is not only considered the norm but is regarded as essential for survival.

It should be remembered that whenever and wherever a successful implementation of an electronic trading project has been completed the commitment of executives has been virtually essential. That commitment cannot and should not be given without an appreciation of the requirements, the implementation plans, the benefits, the critical success factors, the risks, the potential impact on present business processes, and indications as to how the project complements or leads future business strategies. To that cocktail add a whiff of pioneering spirit, the scent of marketing advantage, and top with evident belief in ET to arouse the more adventurous executive's interest. If commitment and vision is not present in senior management then Electronic Trading projects are liable to languish. Without imparting basic knowledge engendering that commitment is difficult if not impossible.

2   THE MODES OF ELECTRONIC TRADING

Electronic Trading solutions both nationally and internationally reveal spectacular diversity. In reality traders need not understand all flavours only the basic principles and be sufficiently aware of the compendium of potential solutions. Informed they can then identify and implement a business methodology that suits them yet does not inhibit the solutions of their trading partners. Electronic data communication embraces the *Electronic Mail, Electronic Data Interchange, Real Time* and *File Transfer Systems* as shown in Figure 4.4.

```
• On-Line Access
      To applications
      To external databases

• Electronic Data Interchange (big EDI)
      Formatted business transactions
      Application-Application processing

• File transfer (little edi)
      Engineering data
      Software
      Data files

• Electronic Mail
      Inter-personal messages
      document/text
```

FIGURE 4.4. Modes of data communication used in electronic trading.

For the executive these modes may be simplified to three basic processes (see Figure 4.5):

- The transfer of information and data between people, referred to as electronic mail (E-MAIL) or, more grandly, Inter-personal messaging.
- The transfer of information and data between a person (P) and an application (A) in real time. This I will refer to as an On-Line Communication System (OLCS).
- The transfer of information and data between applications, which I will term Electronic Data Interchange (EDI).

In an electronic mail environment people are the senders or recipients of information. Senders make decisions about what is sent, when it is sent and to whom at transmission time. Similarly the recipients expect to be able to view their 'in-basket' and decide whether to reply to, bin, store or route for further processing received transmissions. At each step people are in control, interactively utilizing mail facilities as necessary. It should be noted that information can be sent across applications by this process and thus achieve a 'pseudo-EDI' environment. For trading partners new to EDI and unfamiliar with EDI processes the personal control this offers provides comfort during initial implementation phases. It should be stressed that continued human intervention in the flow of EDI transactions is not recommended if organizations are to benefit fully from EDI. It should also be noted that mail systems are usually designed on electronic forwarding principles with expected delivery times pre-determined by 'mail classes' that may not offer the turnaround times required in specific EDI operations (see Figure 4.6).

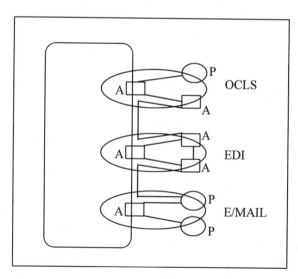

FIGURE 4.5. Simplified electronic trading model.

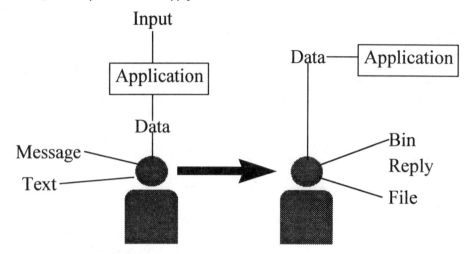

FIGURE 4.6. Electronic mail interactions.

For an On-line Communication system people establish a connection from a work centre terminal to an application. They then interact in real time with that application. The application may be an on-line product catalogue and ordering system through which customers submit orders based on the latest information and get immediate order confirmation. Secondary phases may involve carriers accessing created despatching schedules to determine the size and nature of goods to be transported. The latter then assess vehicle requirements, enter details and trigger the electronic transmission of all transactions associated with the transportation of the goods, these transmissions probably as EDI interchanges. On-line Communication systems have diverse functions and can be located, according to their role, anywhere within a supply chain. Their nature and number depend on the business strategies of trading partners and their competitors.

Probably the most misunderstood of the electronic trading techniques is EDI. In an EDI relationship information is transferred from one application to another. Often overlooked is the importance to an 'EDI' event sequence of On-Line Communication applications or their localized equivalents to initiation of EDI processing. If rubbish is entered then EDI expedites the proliferation of further rubbish unless applications are sophisticated enough to counteract the dangers. When implementing EDI it is important to ensure sufficient consideration is given to the quality of data capture in the 'paperless office'.

The linking of applications is routine Data Processing within a company or organization. For EDI the 'DP system' embraces elements in external organizations. Within an organization the control of database and file structure allows the creation of systems tailored to meet specific needs based on unique blends of operating and networking software.

With EDI associated applications are not within the remit of a single company and 'in-house' structures of the transmission data differ between trading partners (see Figure 4.7). Obviously prior agreement concerning exactly what is transmitted and the subsequent processing is imperative. Thus with EDI the necessity to create standards defining the content and structure of information conveyed between applications. EDI standards define message contents and not communication protocols used in transmission.

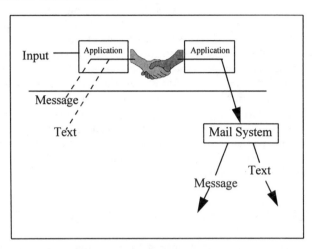

FIGURE 4.7. EDI, applications to applications processing.

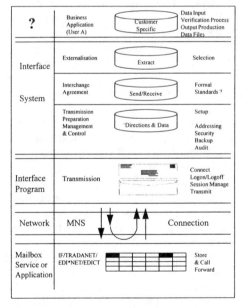

FIGURE 4.8. EDI interfacing functional components.

A typical EDI environment links multiple applications through an 'Interface System' that formats and reformats data to and from agreed standards (see Figure 4.8). It also embodies the management, audit and control functions of the EDI processes. The 'Interface System' is logically independent of the choice of communication protocols and is geared to EDI translation and routing the flow of EDI transactions through communication channels. When trading partners adhere to different EDI standards the Interface Systems must cope with different syntax standards (UN/EDIFACT, UNGTDI, X12), differences in levels of a syntax standard, and differences in messages within a standard syntax. In extreme situations the value of using an EDI standard becomes questionable and reversal to 'little edi' may appeal (using agreed file structures between trading partners not in an accepted EDI standard message format). Finally an interface system selects, which communication protocol, application or EDI Service is appropriate and calls the requisite interface module. The function of the latter is to establish communication links and transmit interchanges. Ideally human intervention only occurs when an EDI interchange is highlighted as an exception condition either by a business application or the 'Interface System'.

3   ESTABLISHING TRADING PARTNER RELATIONSHIPS

The realization amongst trading communities that a move from adversarial to partnership relationships is mutually beneficial prompts a need to re-examine the extent and nature of the ET scene. The 'paperless trading' theme is passing with attention focusing on how ET facilitates improved relationships and workflow with customers, colleagues and suppliers. Communication standardization engenders pressure to widen the ET net and make available to potential associates systems previously restricted. This demand for a free flow of information exposes many nagging issues effecting the ease establishing new trading partner relationships.

The most poignant question posed by many potential exponents of ET is 'how do I know who I can trade with?'. Within Europe associations, consultants, service and software providers all help identify and implement trading partner relationships, however the demand is for a readily available means of identifying potential trading partners. For electronic mail the X500 directory may be the remedy, but for EDI an open directory approach is not appropriate. The EDI concept is movement of information across applications according to pre-determined agreements, therefore a dialogue between trading partners occurs prior to processing. EDI requires an international yellow pages facility allowing traders discover who their potential trading partners are throughout Europe and how to establish an electronic mail link prior to negotiating the EDI relationships.

Extensive study of the legal implications has improved understanding of the concerns needing resolution (see Figure 4.9). International initiatives like the UNCID rules, the CMI, the recommendations of UNCITRAL, the CCC and the Council of Europe and the work of TEDIS and the UN/WP4 working party have all been welcome pan-European developments. The legal issues will not, however, be resolved by discussion but by legislation or case history. Some business communities, reticent about the effects of ET, still use legal and security issues to delay change. The law across Europe needs harmonization to meet the demands of a single market and it needs rapid evolution to meet the demands of the electronic age. All practitioners in IT and ET will be thankful if the efforts of TEDIS 2 and the United Nations Working Party on Facilitation of International Procedures bear fruit.

ET technical implementation issues are more readily resolved with increased availability of simple and relatively low cost solutions. The inhibiting factor for many remains the complexity and duplicity of software perceived as necessary for communication to trading partners on diverse systems using different standards. Hopefully the maturing of UN/EDIFACT standards, the spread of OSI understanding and increased service interconnectivity will alleviate some concerns. Ideally present users will lead by migrating from the present hodge-podge of proprietary standards to EDIFACT.

The fuzziness of the electronic trading models is augmented by the endeavours of communication suppliers to sell solutions based on their present offerings. This contrives to create a confused image on the merits of potential processes and how to apply standards. In application to application environments communications can simulate electronic mail systems. Similarly the basic elements of an EDI Service can be created using electronic mail protocols. With On-Line Communication Systems people may interact with an application that is, in its turn, holding conversation on behalf of the operator with other applications that are creating EDI transactions. Traders need to be honestly informed about the services available to them and achievable performance criteria. The hype surrounding X400, X435, X500, interactive EDI, event driven EDI, fast response EDI, X25, ISDN, Frame Relay, ATM, batch EDI, gateways, CALS and EDI-mail suits consultants and Service providers but is not in the best interest of creating a comprehensible and open environment. Hype and poor advice has initiated inappropriate management decisions in

```
* Terminology
* Protocols
* Standards
* User obligations
* Liability
* Message content
* Confirmation of message content
* Auditability
* System integrity
* Dispute resolution
```

FIGURE 4.9. Concerns in an electronic trading environment.

the past and will do so again. The strategic requirements of the business should be paramount, the system to meet those needs is selected only when the former are clearly defined.

The need for unbiased guidance, education and drive in the electronic trading arena is continuous with pressure mounting for a minimum number of co-ordinated umbrella organizations based geographically or representing specific commercial sectors across Europe. Some business sector organizations, such as EDICON, EDIFICE, ODETTE, CEFIC and the EAN have emerged to promote electronic trading, their success encouraging other business sectors and professions to create new representative bodies such as EDIFICAS and the European Oil and Gas EDI Group. Within national boundaries standard bodies are present but influence varies. In some, like France with EDIFRANCE and Holland with EDIFORUM there is a unified approach. Others, like the UK, who split efforts and created rival standard bodies with diminished effectiveness are now calling for a co-ordinating body or preferably, amalgamation.

These umbrella organizations must be seen to be independent. New partners should not feel restricted to a particular VAN, PTT, supplier, software house or communication protocol, a practice only too evident in many initiatives. The confidence of open system devotees and proponents of cross industry trading relationships wanes each time projects are described as 'cargo traps' or trading community solutions involve 'exclusive' deals. EDI software logic shows distinct and definable interfaces in the processes involved when conveying information across applications. Each offers an opportunity to introduce competitive software, communication protocols or alternative VAN/PTT Services. Some software houses show it can be done and the alternative nightmare battery of dedicated systems for each trading partner or community avoided.

Generally the borderlines between the spheres of influence of the various international standards bodies is sufficiently hazy to warrant intrusions or misuse of the defined standards, and this is extenuated by what appears to be an uncoordinated proliferation of boards, bodies, committees, and interest groups across Europe. In each of the latter there are the expected mix of political elements, some supportive, some wishing to exploit the situation, others delaying or deliberately hindering progress. It is remarkable how far EDI has progressed through the political melee when you consider its potential for social and economic change.

Fortunately this divergence of interests is recognized and an 'Inter-Agency Coalition for EDI' has been endorsed by the Secretary-General of the International Organization of Standardization (ISO), the Executive Secretary of the UN/E CE, the Director of the CCITT and the General Secretary of the International Electrotechnical Commission (IEC). The question remains as to its exact function and its potential effectiveness. It must be careful to avoid being sidelined into academic questions that users find irrelevant and hence sacrifice its potential influence.

4   ADOPTING A UNIFORM STANDARDS POLICY

Successful implementations of Electronic Trading imply that to be effective requires thinking on a global scale across industries with a view to laying the foundations for solving tomorrow's problems today. Approaching an ET project necessitates a visionary who appreciates both present needs and future possibilities. The success of CEFIC EDI project in creating a peer to peer electronic trading relationship across Europe was owed, in some measure, to the willingness of an industrial group to look to the future, adopt ambitious aims and provide the commitment and co-operation needed to succeed. Members of CEFIC established the service level requirements for international and inter-company trading and implemented a policy to meet those requirements. Their identification of the need to encourage the adoption of a single international standard, EDIFACT, and the emerging X400 communications protocol was sensibly balanced by a pragmatic approach to extant projects.

This visioning is necessary to overcome the tendency of people to obviate shortcomings in their home-grown offerings. If conversion to adopting open systems and uniform standards fails then community islands emerge constrained to today's solution of yesterday's problem (or is it yesterday's solution of last week's problem?). If you consider that one of these islands could be the USA with ANSI ASC X12 then you may size the problem. The advocates of universal standards can only be discouraged by the conflicting activities of many national bodies. In the United States the Department of Defence (DOD) issued a policy memorandum officially committing the department to ANSI ASC X12, a move not geared to improve the acceptance of EDIFACT. The EDI Standards Committee of the American Petroleum Institute (PIDX) actively modifies existing ANSI X12 messages and is developing new ones whilst the newly formed European Oil and Gas EDI Group aims to promote EDI in the oil and related industries, to develop UN/EDIFACT based messages for international use and liaize with other groups internationally with a view to developing common standards. The task of the European body is not an enviable one.

The American issues in evolving from ANSI ASC X12 to UN/EDIFACT are appreciable and commitment to a future move to UN/EDIFACT is the best expected, indeed the most fertile ground for implementing international standards are the relatively virgin territories with little or no previous investments to protect. Unfortunately countries with a prior commitment to Electronic Trading naturally wish to protect investments and avoid 'grasping the nettle'. The pressure being exerted on the X12 committee by the Information Systems Standards Board of ANSI may threaten the future of the committee if it fails to provide a viable plan to move towards international standards, but it does not overcome the refusal of the transportation organizations within X12 to adopt UN/EDIFACT. Indeed they want X12 to be the universal standard and resent the 'European' interference.

Each country participating in a pan-European electronic marketplace needs to adopt a policy of harmonization, phasing out local standards and discouraging the emergence of national message dialects. It is encouraging to see the International Article Numbering Association (EAN) reporting increasing use of EANCOM as the EDI standard. The latter, although itself considered an association dialect, is appreciably better than each national body promoting local standards such as TRADACOMS. If countries do not concentrate their efforts towards adopting a uniform standards policy then over complex interface systems will continue to be needed. It was interesting to note the 1992 January issue of the Electronic Trader report on the UK's Article Numbering Association (ANA) explanation of their TRADACOMS Service covering 'TRADACOMS TDI', 'TRADACOMS EANCOM' and 'TRADACOMS UK EDIFACT'. It appears their intention is to 'promote the admirable idea that 'standards are not an issue', and that the ANA can support them all'. This echoes a view attributed to proponents of the IBM Information Network Information Exchange Service, however in that case they advocated a message delivery service independent of message format standards. For the electronic trader standards are a major performance and cost issue, especially if required to purchase and maintain software catering for different message levels within a standard, different flavours of the same messages and an array of different standards.

5   SERVICE INTERCONNECTION

The need to communicate across different Services is contentious. For electronic mail the definition of the X400 Interpersonal Messaging standard by the CCITT created an environment in which objection to inter connection was difficult. This improved the view of electronic mail interconnection globally. In Europe the pace is slow but improving. Thus, despite the complexity off the X400 addressing struct-ure, freedom of choice and single point of entry is encouragingly close.

For EDI interconnection remains an issue. Various interim solutions have been considered prior to the introduction of an X435 solution, especially bridges through Electronic Mail Services making beneficial use of the latter's inter-connectivity. Concerns over security, allocation of responsibility, service levels and charging remain coupled with additional processing requirements to handle the character-istics imposed by the electronic mail bridges. In the US a rather cumbersome X12 mailbag was introduced amid concerns about its use and effect on the timescale for an X435 solution. In Europe an enhanced form of the Odette File Transfer Protocol is being adopted as the immediate solution by some EDI Services. The latter option was always feasible, the real issues being marketing strategy not tech-nical. The absence of acceptable interconnection facilities perceived as protecting a captured customer base with revenue generation from locked in communities. With AT&T, IBM and BT in the UK directly interconnecting the signs are that Service providers are realizing the negative effect of this myopic policy.

Prospects for EDI interconnection in Europe improved further when the European Commission commissioned the creation of a clearing house for all European EDI providers. A consortium of PostGEM, Infonet, Swedish Telephone

International and TSI (a subsidiary of Telefonica Servicios) are to provide a central hub through which all international messages can be routed.

This X435 service is intended as an open system to all European EDI Services and as a catalyst for the adoption of X435 messaging standards against the interim X400 P2 offerings.

6   CONCLUSION: THE PILLARS OF THE EEC

The Common Market is represented has built on three pillars, Political, Social and Economic. To these we must add a fourth, Electronic (see Figure 4.10). The Political, Social and Economic pillars are endangered by the pressures of political intrigue and nationalistic rivalry, but the Electronic Pillar is developing and strengthening as more businesses realize the opportunity and need for co-operation and partnerships across Europe. Electronic Trading is emerging as fundamental in underpinning Europe.

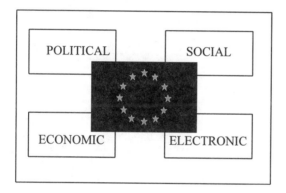

FIGURE 4.10.   The pillars of EEC.

5 Electronic Commerce – A Vital Component in Logistics Strategies

J. Jenkins
Director, Corporate Affairs
GE Information Services

1 INTRODUCTION

The pressure is on to improve inter-company communications, nationally and internationally, across the supply chains in which we all operate. At GE Information Services (GE), we are seeing the emergence of an exciting new phenomenon – what has been called the 'Virtual Organization'.

This is a community of co-operating companies or 'business partners', who have adopted electronic commerce, enabled by technologies such as EDI, to create a closely coupled, highly efficient and integrated supply chain – in effect, many businesses working as though they were one.

The resulting capability for business reconfiguration has been identified by MIT in their 'Management in the 1990s' research programme as a key strategic success factor.

GE is helping companies and supply chains to build and support Virtual Organizations. Its services, such as Tradanet, with Europe's largest electronic commerce community consisting of thousands of companies in the UK, are a fundamental part of a growing, global electronic business community.

This chapter examines the impact of Information Technology (IT) and telecommunications – in particular electronic commerce – on the supply chain.

2 THE VIRTUAL ORGANIZATION

2.1 Management in the 1990s

'Management In The Nineties' is the largest, and most authoritative, research programme yet conducted into the impact of information technology upon organizations and their abilities to survive and prosper in the competitive environment of the 1990s and beyond. The five year, $5 Million, programme was conducted as a partnership between a number of large corporations and the Sloan School of Management at MIT. Some of the key conclusions of the programme were recently published, by Oxford University Press, in 'The Corporation of the 1990s' by Michael S. Scott Morton. GE became involved in the research programme in 1989 as a result of our experience in the application of EDI to business – which had already been identified by MIT as a key enabling technology for 'Business Network Re-Design'.

In order to assess the likely future impact of Information Technology upon businesses, MIT first carried out a number of detailed studies of major techno-logical innovations of the 1980s; the resulting business benefits; their impact upon competitive advantage; and the success, or otherwise, of the participants.

From this work they were able to draw a number of important conclusions concerning the impact which IT will have upon businesses, and to construct models which better enable the strategist to visualize and direct the resulting changes.

> *The fundamental premise is that it is no longer a question of whether IT has a strategic role but of how to exploit IT in the strategic management of the business. How is it possible to identify the strategic IT applications, to reconfigure the business to exploit IT capabilities and to use IT to differentiate from your competitors?*

MIT classify such changes in terms of a hierarchy of five levels of business reconfigurations:

2.2   Localized exploitation

This is IT implementation within a business function within an organization, typi-cally to improve the efficiency of a particular task. The installation of a PC based, purchase ledger or customer database system for example.

The benefits at this lower level are localized, giving greater efficiencies in narrow areas of the business.

2.3   Internal integration

This level extends Localized Exploitation by integrating key internal applications to establish a common IT platform for the business – by means of local area networks or corporate databases for example. Typical applications are integrated sales forecasting and production scheduling, or integrated ordering and warehouse systems.

The benefits of this approach are, typically, greater efficiency improvements, information sharing and responsiveness to customer demand.

These first two levels are evolutionary, requiring relatively small changes to the business processes. In contrast the remaining three levels are revolutionary – requiring radical changes in business practices. At levels one and two IT is an administrative resource, used to support existing business strategies. At the higher levels IT is a strategic investment which enables new business strategies.

2.4 Business process redesign

The central premise at this level is that IT is a platform for designing business processes and that it should not be simply overlaid on the existing organization. Now it is possible to start restructuring business processes and aligning the IT infrastructure with them.

2.5 Business network redesign

Thus far we have seen IT-induced reconfiguration within a single organization. Significantly, this level represents the use of IT for redesigning the nature of exchange among multiple participants in a business network, changing the boundaries between one organization and another and thereby creating a more effective business network.

The key IT enablers are communication network based and include EPOS, EFT, E-Mail, Database Enquiry and exploitation of EDI; i.e. the 'Electronic Commerce Services' which form GE's product portfolio. Example strategies that result from this level include QR (Quick Response), ER (Efficient Replenishment), CALS (Computer Acquisition and Logistics Support) and other supply chain management strategies. These provide enhanced operational efficiency, more effective market positioning and greater opportunities for business partnerships. This level of the model corresponds directly to the Supply Chain Integration approach now being adopted by Tesco, and to the AA's relationship with its panel of BROKERNEI insurers. For such strategies to succeed it is essential that the potential benefits are fully understood by all members of tbe supply chain (or Business Network in MIT terminology).

2.6 Business scope redefinition

The highest level represents the enlargement or shifting of the business scope by the substitution of traditional capabilities with IT-enabled skills, for example EDI, Systems Integration and Information Databases.

Effectively, level five offers the potential for redefining the business. A classic early example of this trend was American Airlines' seat reservation system, SABRE. Enabled by information databases, EDl and telecommunication networks, SABRE was transformed from an internal seat reservation system supporting the sales of American Airlines own capacity, to a service for American's industrial competitors – where it developed into the most profitable unit of American Airlines! Good UK examples of business scope redefinition, enabled by EDI, are already emerging, but perhaps the most obvious example was that of International Network Services (INS), set-up as an ICL/GE Information Services Joint Venture specifically to market EDI and other electronic commerce services.

The benefits at this level are, therefore, highly strategic and provide opportunities for growth, added value and new businesses. Almost by definition, it is very difficult to anticipate such developments – but the MIT results can certainly help in the identification and evaluation of opportunities.

2.7   Redesigning the business network

There is no sustainable competitive advantage to be gained from Information Technology. The pace of technological development means that innovative IT solutions can inevitably be reproduced, or substituted, at lower cost by responsive competitors. This fairly obvious conclusion is, however, not the whole truth as identified by the MIT programme. Technological innovation does allow an organization to create temporary enhanced competitive position, which may then be exploited to redesign processes and establish new types of business relationship which then prove extremely difficult for competitors to substitute or copy. Thus, by continual IT-based innovation, very strong market positions can be created. A very good example of this approach can be found in the ASAP system, implemented by Baxters in the US Medical supplies market. This is particularly true where EDI and other electronic commerce services are employed, and then exploited, to integrate, re-position, and even eliminate, business processes between organizations within the 'Business Network' (or supply chain). It follows that the type of business relationship adopted with key members of your business network must be considered alongside the role of IT integration.

The following model illustrates the strategic options for business network re-design, with regard to these two basic considerations:

- **Electronic Infrastructure:** The position currently adopted by the majority of users of TRADANET, and other EDI services, is characterized by the use of public data standards and a common role for IT; with loosely coupled business relationships. Although there may be some barriers to entry for new participants there is little room for new advantage among exsting players.
- **Competitive Advantage:** By employing proprietary linkage, or data structures, it is possible for the 'first movers' to gain some (albeit very short term) competitive advantage or supplier lock-in. This position is almost impossible to sustain and is increasingly unacceptable to business partners or 'spokes'.
- **Collaborative Advantage:** This is the position to which several of the more enlightened UK EDI 'Hubs' are moving, with certain of their key suppliers. Whilst still employing common Electronic Commerce enabling technology they are forming commercial relationships which allow the technology to be exploited for mutual collaborative advantage. This may be accomplished by removing redundant, or overlapping, processes between the trading partners – e.g. through the use of self-billing, exchange of forecasting data, account reconciliation.
- **Business Network Redesign:** In this position the IT infrastructure is exploited to strengthen and modify the nature of business relationships

between members of a 'Business Network'. Experience shows that the 'unique role' of information technology should be based upon 'added value' applications and/or functionality rather than proprietary links or standards if this position is to be sustained. Proprietary (lock-in!) IT solutions will be strongly resisted by business partners and will, in any event, be overtaken by the more rapid and cost-effective development of open solutions.

With a closely coupled business relationship it will be possible to maintain the unique nature of this IT 'added value' through sustained innovation. In this position it is possible to develop business relationships which achieve a significant level of sustainable competitive advantage.

2.8 Towards the virtual organization

The MIT programme contains a considerable volume of material which identifies and explains the key strategic considerations in Business Network Redesign; examines the roles of the business network and relates both of these to the potential business benefits. Some of the analysis is drawn directly from our experience with TRADANET and other electronic commerce services and we are very keen to enable our customers to benefit from this work. Whilst the implementation of EDI for Electronic Commerce (or **'doing business electronically'**)suggests benefits arising from administration and inventory, it is important to recognize that process linkages and information/knowledge sharing up and down the supply chain can ultimately yield far more sustainable competitive advantage.

The following model summarizes the various roles of Electronic Commerce or, more generally, electronic integration in business network redesign. MIT conclude that these roles are hierarchical – offering increasing levels of strategic capability, or opportunity for sustainable competitive advantage.

The roles are categorized as:

- **Transactions:** The use of EDI to exchange the equivalent of paper trans-actions – e.g. order and invoices. The benefits are largely in operational efficiency, accuracy and reduced administration costs. There is some market positioning advantage in being a first mover and hence, potentially, able to create or influence the industry standard. There is little reason to be selective in terms of trading partners and the longer term strategic impact is likely to be low.
- **Inventory:** The electronic trading relationship(s) makes inventory 'available and visible' from one party to another. Status information is enhanced and the movement of goods is triggered – e.g. enabling JIT manufacturing. Inventory tends to be forced back upstream and so the resulting benefits drift downstream. There are significant potential benefits to the consumer through the progressive reduction of inventory and/or work in progress

(thus cost) from the supply chain. The shifting of costs will generally need to be reflected in the contractual relationships – or trading discounts.

This role of electronic integration is not generally relevant in financial services, for example, where there is no physical inventory. This might help to explain the comparatively slow adoption of EDI in this sector.

- **Process:** Specific trading partners integrate their business processes through electronic links, removing redundant or overlapping processes to create a new and more efficient supply chain or 'integrated business network'. This type of relationship will require a 'partnership' approach and will be adopted only with key members of your business network. The MIT case study on GE Lighting is a classic example of this role of electronic integration in the retail industry.

- **Expertise:** In this role skills and expertise are shared among the members of a business network. Very close, and specialized, commercial relationships will be required and the opportunity exists for cooperating organizations to establish sustainable competitive advantage. A UK example would be the relationship between Courtaulds and Marks & Spencer for clothing design. Using both CAD/CAM and EDI technology; retailing, design and manufacturing expertise are effectively being shared electronically in a unique relationship.

It is important to recognize that GE, as a leading supplier of Electronic Commerce Services, may already be positioned as an important partner in your supply chain or 'Business Network'.

The models outlined above apply no less to the relationship between GE and our customers than to that between a major retailer and its suppliers. The final role of electronic integration – that of sharing skills and expertise – is absolutely appropriate to this relationship and offers the very highest level of potential strategic capability to our customers.

GE's business lies in the provision of productivity solutions based upon Electronic Commerce Services which enable our customers to develop and exploit these integrated supply chains – or Virtual Organizations.

3   EVOLUTION OF THE VIRTUAL ORGANIZATION

During the past 12 months we have seen rapid acceleration in the growth of our electronic commerce communities – on average over 100 new companies each month have joined TRADANET. But not only are the total numbers growing rapidly, the exploitation of EDI and the development of our customers' business networks is significant. The fundamental difference with the user communities of GE compared to our competitors – even other countries – is the reality of progress, the reality of change and the depth of activity.

The use of EDI as a foundation for change is firmly established across a huge section of commerce and industry.

Here are some statistics regarding our EDI communities today:

40,000 users worldwide
7200+ users of TRADANET
1500+ users in managed communities
35 market sectors
78 of the Times Top 100 companies
9 of the top ten Retailers
70 of the Times Top 100 Exporters
5 of the top 6 Vehicle Leasing companies
All major Clearing Banks.

These statistics, highlighting some aspects of the GE community, bear witness to the depth of activity.

There is no longer a need to describe a company's plans, nor a company's targets for the future – one merely has to examine the historic activity of many communities to see the benefit and change GE is bringing to business today.

We are seeing a series of dramatic changes; changes that vary from industry to industry – but changes that have a common underlying theme. Companies are forming closely coupled, highly efficient communities of businesses using technology and services for all intercompany communications, using the technology infrastructure of TRADANET to reshape the boundaries between companies, resulting in the formation of business networks with cost taken out to the benefit of all parties.

Here are just a few examples.

Retail has been one of the strongest and most effective exponents of EDI since the pioneering work of the ANA in the late 1970s and early 1980s.

Tesco – probably Europe's leading exponent of EDI and Quick Response – has over 1350 trading partners exchanging an increasing variety of business documents electronically.

An example of the erosion of boundaries in retail can be found in the exchange of forecast information. This is being undertaken by Tesco for both short and long life products and results in suppliers having the ability to plan their production more efficiently, to reduce stock holdings, to increase service levels and to eliminate obsolescent stock.

The net result is members of the business network working to accurate and timely information, taking cost out of their joint operations.

Tesco is one example but almost all of retail is active in EDI. Such is the pace of change now, and such are the products and services available to facilitate this change, that companies can make huge inroads to the formation of their virtual organizations in remarkably short timescales.

Superdrug, for example, set themselves a target of 12 months to establish a community of 100 trading partners on TRADANET. Working with GE in a roll-out program with products like INTERCEPT-PLUS, they achieved this in 6 months. This community represents over 80% of Superdrug's orders by value. Somerfield, a major in UK Food retail, rolled-out their supplier community of 400 suppliers on TRADANET in just four months – one of the fastest community developments in Europe.

Such is the acceptance and penetration of EDI and Efficient Replenishment in UK retail that anybody not involved is increasingly at a severe competitive disadvantage.

The principles of Quick Response and Efficient Replenishment are not limited to retail only. In Manufacturing they call it 'Just In Time' (JIT).

The Electronics industry is another significant user of TRADANET with companies such as ICL, NCR, GPI, MITEL and SONY.

Sony's Bridgend manufacturing plant produces one and a quarter million televisions a year. Their implementation of EDI has been focused on supporting their JIT strategy. As a result, Sony now order over 750 components via EDI representing over 50% of orders placed with local suppliers.

Clearly, as in Quick Response, the focus of Just In Time has been to reduce lead times. EDI has made a significant contribution to Sony's ability to reduce their order to delivery cycle by approximately two thirds.

The Banks have an important role to play and GE is committed to working with our community and these financial institutions to expand the financial services available via EDI.

Finally, Government. Whilst there are some interesting initiatives in local and central government – particularly the activity of HMSO – government must be one of the biggest potential users of EDI. The organizational dynamics are different but the potential for change and advantage is equal if not greater, than many areas of the private sector.

In summary, there is a wealth of EDI and Electronic Commerce activity in GE's communities.

This activity is strategically focused on the principles of bringing improvements to the interaction of companies in their changing business networks.

This activity is facilitated by a partnership approach and by the availability of the knowledge, experience and products and services – our Electronic Commerce Services – that we deliver to our customers.

4   ELECTRONIC COMMERCE SERVICES

4.1   GE information services – in a class of its own

We have discussed the importance of Electronic Commerce Services and we have seen just how succesful GE has been at developing them.

Why is this the case? Why is GE different? Why have we been so much more successful than our competitors? The reality is that we take a very different view of electronic commerce SERVICE from any of our, supposed, competitors. Of course we need network carrying and data switching capacity, but successful EDI exploitation and business network change is heavily dependent upon the quality and availability of 'added value' services:

- Guaranteed service availability and reliability
- Security and audit facilities
- The widest possible range of user connectivity options
- Highly functional and reliable user software
- Constant Help Desk availability
- User installation and training services
- Consultancy and project management
- Marketing and community building services
- International 'roll-out' management.

It is the surety of service offering and the 'know-how' reflected in this 'added value' which sets GE apart.

4.2 Supporting the virtual organization

We have already described the considerable activity and accelerating growth in our customers' increasingly virtual organizations. These are fundamentally reliant upon products and services – solutions – that GE delivers: our range of electronic commerce services – software, delivery services, industry-focused applications, people-based services and global support.

The foundation – the bedrock – of our service in the UK is TRADANET, probably one of the world's most sophisticated electronic commerce services.

'EDI and electronic commerce is not a choice .... (it is) the inevitable way business will be done'. Massachusetts Institute of Technology *'Management in the 1990s Programme'*.

4.2.1 TRADANET

TRADANET supports Europe's largest electronic commerce community. With its national and international connectivity, it offers its users a range of features, facilities and services unmatched by any competitive offering in the world. Here are some of them:

- **High Performance:** TRADANET offers the highest performance EDI service available, providing instantaneous transfer of data from sender to recipient within the service.
- **Features Rich:** TRADANET offers the user a high degree of functionality, allowing the service to be managed and controlled according to the business requirements. Here are just a few of the comprehensive set of facilities available:
  - Detailed historic audit of mailbox information, available on-line.
  - Detailed historic audit of postbox information, available on-line.

- Comprehensive audit trails for each telecommunications session.
- User-controlled trading relationships set-up and the availability of on-line audit report.
- Sophisticated data selection and extraction capability.
- User controlled file management facilities.

Users of TRADANET are free to set up new trading relationships, for any standard documents, as their business requires. This they do under their own control.

- **Around the Clock Availability:** Availability of TRADANET is maintained at the highest possible level on a 24 hour, 365 day basis. (Over 99.75%, including scheduled outages, for 1995.)

  This is very important for national trade, but vital for international trading between different time zones.

- **Around the Clock Service:** Full 'hotline' Service Desk facilities are available around-the-clock.

- **Resilient:** The unique architecture of the service is based upon dis- tributed processing nodes that provide effective back up for each other. As one node is removed from the service through failure or more commonly for mainten- ance purposes, a second 'hot standby' can be switched in to take over. This is a regularly tested feature of the service.

- **Disaster Proof:** Not only are back-up processing nodes available on the primary processing site, a second site, interconnected to the primary site by high speed land line and microwave links, is available in 'hot standby' mode.

- **Safe and Secure:** The service incorporates a large number of security and data integrity facilities that ensure total confidentiality and privacy of user data.

  TRADANET does not inspect the data it processes in any intelligent way. It is only concerned with the envelope details – not the contents of the envelope. This is a primary user security requirement.

  The security aspects of the service are independently audited by one of the world's leading experts in security of telecommunication and computer based systems. Such audits are carried out on an annual basis.

- **Dedicated Processing Facilities:** For maximum security the computer facilities responsible for processing the service are dedicated to the service. No other application processing is permitted.

  This is a unique feature of our service.

- **Highly Flexible Communications Access:** The network facilities offered by GE provide users with a very high degree of flexibility of connection from their computer systems. The service offers a wide range of communi- cations protocol support – SNA, Async MNP (block and stream mode), X.25, OFTP and X.400 – which is second to none in the UK. Additionally, line speeds from 1200bps dial-up connection to 64Kbps access through dedicated kilostream lines are available according to our user's business requirements.

The 'Mercury 5000' carrier network is developed to OSI standards and is one of the largest and most sophisticated privately owned X.25 networks in the UK. It provides local-dial access to over 95% of the UK business population.

Highly flexible commercial arrangements are available to service users, including a single worldwide contract, single price list and consolidated billing.

- **The International Service:** TRADANET is part of GE's global community of 40,000 companies – the largest electronic commerce community in the world. As such it is a truly international service. The end-to-end audit trail facilities which integrate TRADANET with our global EDI service, EDI*Express, are unique in that no other inter-network connection in the world can offer similar features. TRADANET's international service spans 100 countries worldwide.

  What's more, the local language support facilities provided by GE Information Services, in 35 centres around the world, form a unique INTERNATIONAL SERVICE CAPABILITY able to build and grow international electronic commerce communities.

- **Standards Support:** GE has constantly taken the lead in development and support of national, and international data standards. TRADANET is well known for its support of TRADACOMS, but we were the first in the UK to provide full EDIFACT support, and GE is actively involved in the development of EDIFACT standards in a number of areas. In the UK insurance industry, in particular, GE has helped to create new standards which have now been adopted by the ABI (Association of British Insurers). Another example, we have given advice and guidance to the ACLM (Association of Contact Lens Manufacturers) in the Optical Industry.

4.3 Commerce*Express services

However, TRADANET is but one of GE's services. We market a comprehensive range of products and services, combined with people-based services, consultancy and training, for example, that enable GE to package up business productivity solutions for our customers – to enable them to develop their Virtual Organizations. This portfolio is branded **Commerce*Express Services.**

Commerce*Express Services consists of:

- **Reliable Delivery Services** – global EDI services, such as TRADANET, E-Mail, X.400 and Information Management Services with the very highest levels of performance, service availability, data integrity, and system security – designed to support business critical applications.
- **High Quality Software Products** – a full range of electronic commerce software across all major hardware environments – from the desktop to the mainframe. Network independent EDI enabling software, capable of handling multiple standards, multiple trading partners and multiple

business processes. Integrated workstation software for supply chain management, financial reconciliation and many other applications.

- **Community Applications and Information Services** – network-based applications for electronic commerce, including POS-based forecasting for vendor managed inventory, electronic payment systems, global reporting, risk exposure management, cargo tracking and status reporting. Access to electronic catalogues, bulletin boards and a wide variety of news and market information services.
- **People Value-Added Services** – the experience and expertise to build and support electronic commerce communities, providing business process consulting, implementation, application integration, training and community development.
- **Global Support** – Commerce*Express Services is backed by full local language support from service centres in nearly 40 countries around the world.

For more information contact:

GE Information Services
1-3 Station Road
Sunbury-On-Thames
Middlesex TW16 6SB

Tel.   44-(0)1932-776000
Fax: 44-(0)1932-776020

6 Global Networking: Why a Managed Solution is Best

G. McArthur
British Telecommunications plc.

1   INTRODUCTION

Michael Porter in his book *Competitive Advantage* states 'information systems [are] having a profound impact on competition and competitive advantage because of the pervasive role of information in the value chain'. Consequently those companies who can link their information systems and communicate critical business information quickly and efficiently should achieve a competitive advantage.

At the centre of this flow of information is a network, with the prevalence of IT systems these networks are data dominated rather than voice. Increasingly as companies move from a national to an international operation their systems and consequently their networks also need to grow internationally. This chapter examines the alternatives available and illustrates why outsourcing the day to day operation of a data network will both improve the service and reduce the costs of operation.

2   WHY HAVE A DATA NETWORK?

Linking your operation worldwide can enhance both your financial performance and improve the service you give to your customers. The benefits a data network can give you are given in the following sub-sections.

2.1   Improved business efficiency

A data network allows you to cost effectively exploit and share the IT resource within your operation, no longer do you need to buy a new system when you open a new office. With a worldwide operation different time zones allow an even more efficient use of systems.

Additionally a data network removes the time delays inherent in many processes through performing transactions faster and more efficiently using Just in Time and EDI.

Connecting your systems together allows for more coordination of interdependent operations and better tracking, easier collation of information which becomes instantly accessible worldwide.

2.2   Improved customer service

Increasing focus on customer service requires a continual improvement in the time it takes to respond to customer enquiries. A network allows immediate access to the systems that can provide an answer and can allow you to be proactive in your customer liaison.

Additionally through automation the quality of your operation is improved, information sent from one system to another electronically does not have errors, whereas 50% of all documents that are re-keyed do.

Clearly a data network is critical to any organization which is operating worldwide and has a large flow of information. So the question we must now ask is what are the alternatives?

3   OUTSOURCE OR DO IT YOURSELF?

When deciding to commission a data network many organizations look to see whether they should build their own or outsource the supply to a Managed Data Network Supplier. Increasingly the trend is away from DIY to MDNS, the reasons are given below.

3.1   Controlling the costs and risks of networking

The strongest argument why to outsource is simply one of cost, a typical network will be operational for 5 years. Over this period with all the costs taken into account a managed solution will be 30% cheaper than a DIY solution. With the ability to buy fixed price contracts there is security in understanding the costs over the 5 years.

In a time of limited capital spend a managed solution has significant advantage. Instead of a high upfront cost as in DIY through the need to purchase equipment, a managed solution has a low connection charge typically 20% of the DIY commissioning costs.

As a managed solution is inclusive of management and maintenance then there is no need to employ staff to manage the network thus reducing people overheads.

3.2   Flexibility

In an environment of constant change with sites expanding or contracting one needs to have a network that is highly flexible. This requirement covers both the ability to cope with traffic peaks and bring on or close sites. As a managed solution is pre-provided and has thousands of other customers then traffic peaks are easily accommodated and new sites can be brought on quickly as the equipment is already on the ground.

Increasingly technology is changing at an ever faster speed, with a managed solution these changes are introduced once they are proven. In a DIY scenario they are expensive to introduce and timing can often be difficult to judge.

3.3 Concentrate on core activities

More and more often Communications Managers are expected to play a strategic role in improving the effectiveness of the business. To be able to do this by looking at the business and its needs a Comms. Manager has to have the time. If they are bogged down in the day to day tasks of running a network then this can be difficult. A managed solution frees the Comms. Manager from these day to day routines to concentrate on the more interesting areas of using technology for competitive advantage.

4   CONCLUSION

BT Global Data Services is one of the world's largest managed data network service. Today it operates in over one thousand cities in 45 countries, by 1997 it will operate in 60 countries. In the UK 80% of the Times 100 use the service, in North America 50% of the Fortune 500.

Our customers come from all sectors and their common need is a global supplier who can provide them with a high quality data network.

7   In-house or Third Party Distribution?

J. A. Sturrock
Christian Salvesen Distribution Ltd.

1   INTRODUCTION

The reasons given for manufacturers, wholesalers and retailing companies contracting out their distribution are many and varied but most are well known to all who operate in this field. After all, distribution contractors like Christian Salvesen, Tibbett & Britten, Excel and Hays have been preaching them loud and clear for the last ten years. They hold out the promise of, among other benefits:

- Incredible Savings on Operating Costs
- Liberating Vast Amounts of Capital
- Dramatic Service Improvements
- An End to Industrial Relations Hassle
- Freedom to Concentrate on the Real Business etc.

This entire chapter could be devoted to promoting these claims, but that would not be very fruitful in terms of understanding the reality. However, if anyone has not had the full sales pitch they need only contact one of the above named firms.

On the other hand there is a considerable, sometimes vociferous, body of opinion that holds that third party contract distribution is a rip-off, involving loss of control, inflexible service and constant hassle. Where does the truth lie?

A study by PE Consulting in 1991 looked at levels of satisfaction among firms who had contracted out their distribution and, while I do not intend to go into the details here, two of their reported findings are relevant to this subject.

2   REASONS FOR CONTRACTING OUT

The first finding was that the most important reasons given overall for contracting out were to Reduce Cost and to Improve Service, although it is interesting to note that within the sub set of grocery retailers, Require Specialist Management was given as the most important.

It is fair to say that the grocery retailers are a special case since their move to contractorization came as a consequence of the even more fundamental decision to go to centralized distribution. Thus they were setting up completely new logistics networks from scratch, and simply had not the resources to finance staff and manage these themselves within the timescales they wished to achieve. Manufacturers and wholesalers on the other hand already have in-house distribution resources available so, although they may wish to restructure these drastically, the choice of staying in-house or contracting out is more finely balanced.

It is reasonable to suppose that the motivation for contracting out depends on the particular situation of the company and I find one useful way of looking at this is on a two dimensional matrix as shown in Figure 7.1. The vertical axis is a measure of the criticality of the distribution operation to the overall success of the firm. High criticality may be caused by the significance of the distribution costs in the selling price of the product, by the need for immediate availability and quick response, or by the need for special handling or environment control. Examples of each of these might be bottled water, fashion goods and blood plasma.

The horizontal axis is a measure of the complexity of the logistics operation. Is it straightforward, requiring standard facilities and labour and working to normal tolerances? Or does it call for highly specialized equipment and skilled personnel performing complex tasks to a high degree of accuracy and reliability? Of course the two axes are not totally independent, as complexity and criticality often go hand in hand, and it is difficult to envisage an ongoing situation where distribution is at the same time complex and yet only incidental to the firms success. So the bottom right quadrant can be labelled 'not applicable'.

In the upper left quadrant I am suggesting that for those companies whose distribution is critical to their success, but at the same time is not too complex, then in-house could well be the favoured route. This should give them maximum control, but without taking on too onerous a burden or requiring expertise not already available from within.

Then there are two distinct situations which would appear to favour contracting out: at the bottom left, where distribution is non-critical and not very

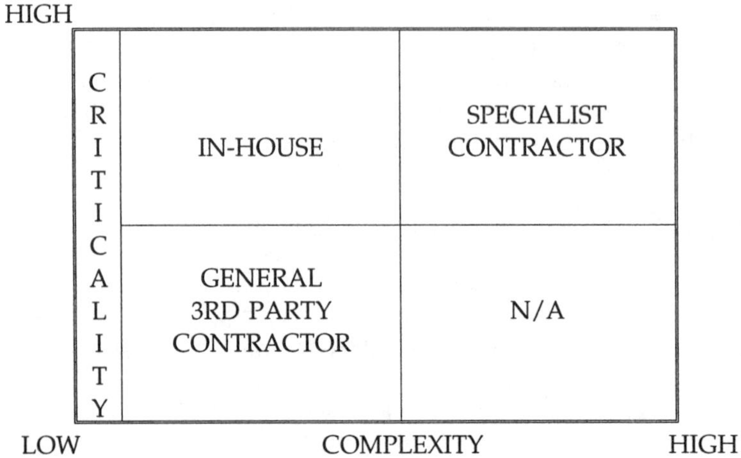

HIGH

C R I T I	IN-HOUSE	SPECIALIST CONTRACTOR
C A L I T Y	GENERAL 3RD PARTY CONTRACTOR	N/A

LOW   COMPLEXITY   HIGH

FIGURE 7.1. Contracting matrix.

complicated there will be a tendency to farm out a mundane task to general contractors who will do an adequate job. The choice of contractor will be based primarily on price and lack of hassle. At the other end of the scale, however, the top right hand quadrant represents logistics tasks that are both critical and difficult. I believe this is the area where the Specialist Contractor should be able to justify his existence. His special skills are needed, not just to meet the immediate require- ments for extreme reliability and service, but to contribute to and adapt in pace with the client's evolving supply chain strategy. Contracts are likely to be of a longer duration, and while price will be one key determinant (as it always is) quality and reliability of service, flexibility and other less tangible factors will be equally important.

This is only a crude analysis which could be expanded into more dimensions, particularly if criticality was resolved into separate cost and service axes. Such elaboration may be left to the academics and consultants but this simple model provides one framework against which to place the success or otherwise of contracting out in different situations.

3   RESULTS OF CONTRACTING OUT

The second point taken from the PE study is that roughly two thirds of those companies who contracted out were satisfied with the results, but one third were not totally satisfied and nearly one in ten were positively dissatisfied.

The first part, that two out of three felt that they had achieved their aims in contracting out is itself a considerable endorsement for the product. However the very significant level of dissatisfaction, either partial or total, is a grave cause for concern. At the same time no one should get complacent about the satisfied two thirds because it may be that, although they have achieved their targets, they should have been achieving much more.

What were the reasons for dissatisfaction? The two most important problems were quoted as Service Failures and Lack of Adequate Management Information. These are just the sort of concerns that one would expect to find in critical/complex situations and suggest that the most likely quadrant for dis- satisfaction on the simple matrix would be the top right. As this is the area in which most of the modern, ambitious contractors have tried to position themselves, it is clearly necessary for their future prosperity to come up with convincing answers to these problems.

4   INTEGRATED SUPPLY CHAIN PARTNERSHIPS

Instances of dissatisfaction with third party contract distribution can broadly be put down to one or a combination of four principal causes:

- wrong decision to contract out rather than retain in-house;

- incompetence on the part of the contractor;
- inadequate, inflexible or inaccurate specification; and
- lack of true partnership relations between contractor and client.

I would argue that the first three of these causes can be avoided by correcting the fourth. In other words the answer lies in developing close partnership relations.

A good contractor, intent on working in true partnership spirit with his clients will point out if a particular distribution task would be better carried out in-house. One may question whether any contractor would ever forgo a potential fee opportunity, but it has to be remembered that most of the respectable contractors are in this business for the long run and know that it is just not worth conning a gullible client into a contract if it is likely to prove to have been wrong in the future.

A good client, looking for a partnership relation with a potential contractor will get to know him pretty thoroughly before making any major commitment, so that the level of competence can be clearly established at the outset.

A good contractor and client will jointly work out the specification and control procedures and, perhaps more importantly, develop a mutually agreed approach to handling the inevitable variations from the specification and identify any areas of uncertainty which cannot be pinned down until later.

The essential message of this chapter then is that the key to success in contracting out distribution operations, particularly in those situations where distribution is both complicated and critical to the client's competitive edge, is the formation of integrated logistics partnerships. It is therefore appropriate to talk a little about how such partnership relations may be established and maintained.

5   ESSENTIAL INGREDIENTS

The essence of any partnership is firstly that the members share a common aim, but each has a different contribution to make towards achieving that aim. Secondly, each member must trust the other to make their specialist contribution in the most effective way and to subjugate their individual short term interests to the achievement of that common aim. Finally there is a continual need for honest and open communication between the partners, so that information about the common aim, the overall performance of the supply chain and the internal and external constraints affecting each member are communicated throughout.

So three of the essential ingredients for a successful logistics partnership are Common Aim, Trust, and Information. Of course each party also has to perform its appointed part of the task, but that should go without saying. A fourth essential ingredient however is Long Term Commitment. Partnerships are not created instantaneously on the signing of an agreement; they evolve and develop over time. This long term characteristic gives the members a degree of security which breeds the confidence to invest their financial and human resources in the partnership. Time also allows the members to develop deeper understanding of each other's cultures, needs, capabilities and frailties.

6   ESTABLISHING PARTNERSHIPS

There are four key stages in establishing a successful integrated logistics partnership. They are:

- Defining Roles and Relationships
- Choosing Partners
- Setting the Ground Rules
- Netting In.

One party has to initiate the partnership, and that will almost inevitably be the client, either manufacturer or retailer, although they may be prompted by consultants or by the persuasiveness of a contractor's salesman. The client best knows what the total logistics task is and can make a preliminary judgment as to how that task should be divided between the prospective partners.

In choosing a partner, the more obvious factors such as competence and financial stability can be fairly easily ascertained from track record, references and published accounts. What is much more difficult is to find out whether there is a real congruence of aims and compatibility of cultures, likely to lead to the development of trust and open communication and whether the partner has the flexibility and responsiveness needed to ensure the long term prosperity of the partnership. At the same time the initiating party should be honestly examining his own organization to see if it also has these necessary characteristics.

But partnerships are not created from cold. Rather they have evolved between partners who have got used to working with each other over a period of years. To quote one example from my own experience, Christian Salvesen began working for Marks & Spencer in 1972, distributing 50,000 cases a year of frozen food to 12 stores. Over 20 years that relationship blossomed into a multi-million pound business for Salvesen, taking them into chilled and ambient foods, and subsequently, largest of all, garment distribution including servicing the flagship Marble Arch store. At the same time, the partnership provided a dependable, innovative and responsive channel for a very significant proportion of M&S's merchandize, a benchmark for other contractor's operations and Salvesen were invited, indeed required, to contribute to M&S's thinking on Supply Chain Strategy.

Even where the prospective partners are well known to each other, the normal format for the selection process today is competitive tendering. However, since modern supply-chain partnerships are very different from traditional trading relationships, I do not believe that this is necessarily the best way. The client is looking for the partner with the greatest potential for contributing to the business over the long term rather than the lowest price for the initial – probably transitory – task. Clearly there has to be some method of evaluation and selection, but this may better be done through a 'consultative' tendering process, in which the prospective bidders are involved with the client in the preparation of the specification or scope of work against which they will bid. Too much emphasis should not be placed on the short term bid price to the exclusion of long term cost effectiveness, and during the bid process contacts need to be established at several different levels of the organizations in order to determine not only operating competence, but also strategic thinking and corporate culture. At the same time it is inevitable that personal chemistry will play a significant part, and that is as it should be as it is individuals working together that will make the ensuing partnership effective.

Next comes Setting the Ground Rules. The first, but rarely articulated question is 'who leads?' While the theme of this chapter is partnership, it would be naive to pretend that all the partners are equal. He who pays the piper calls the tune and perhaps the readiness of distribution contractors to accept the sub-servient role and in effect do as they are bidden, has allowed them to form better partnerships than have yet emerged between manufacturers and retailers, where each is struggling for power.

Points to be covered in setting the ground rules of a partnership are:

- Scope of Work
- Division of Responsibilities
- Required Service Levels
- Dedicated or Multi-user Facilities
- Performance Monitoring
- Open or Closed Book Accounting
- Ownership of Assets
- Length of Commitment
- Exit Arrangements.

These should all have been covered in the tendering process and most are self explanatory. However, a few deserve special comment as follows:

- My experience has been that open book accounting is more conducive to true partnership working than closed book or fixed price, provided that the parties can agree at the outset on a level of profit for the contractor, commensurate with the financial and human resources he is investing in the operation, but also having regard to the client's trading position. Under a fixed price deal, the contractor's first concern will inevitably be how he can protect his profit, rather than how he can enhance the cost effectiveness of his service to the client. Open book can be modified to include elements of sharing cost savings and over-spends so as to give a more immediate incentive and some sharing of interests to help his client become more cost effective so that he will grow and enable the contractor to grow with him.
- Length of Commitment, Ownership and Exit Arrangements are all interlinked. In an open partnership, the decision as to who owns the assets should be based on which partner can obtain the cheapest finance for these.

If the client owns them, it may appear easier for him to change contractors in the future, but in practice, where the contractor owns the assets the exit arrangements will include options for transferring these on termination. It is important that exit arrangements are properly covered at the outset, because over a long term partnership circumstances will change and the partners' interests may begin to diverge. I believe it is easier to commit more wholeheartedly to a partnership if you know that a fair and equitable divorce procedure is available should you wish to part in the future.

- Should these ground rules be enshrined in a contract? I am not a great believer in legally binding agreements, because I do not see how every eventuality can be taken into consideration when drafting a contract to last five or ten years. If the parties end up suing each other in the courts the partnership has irrevocably broken down anyway and the only real gainers will be the lawyers. In fact many of our longest partnerships operate on Memoranda of Understanding rather than legal contracts. While outlining the rights and obligations of both parties, these encapsulate the spirit rather than the letter of the agreement.
- The final phase I call 'netting in', where all the parties get tuned in to the same wavelength. The role of Information Systems and EDI has been the subject of innumerable books and papers, so suffice it to say that they provide the spinal chord, or central nervous system of any partnership, permitting the rapid and accurate interchange of information. However, it is not enough to have all the high tech systems installed and running. There is an equally strong need for the people at all levels of the two partner organizations to get to know each others' procedures and styles. Time spent in pre-operational planning, training and interchanging staff is an essential investment if the combined operation is to get off to a smooth start from day one.

7 WORKING IN PARTNERSHIP

Establishing the partnership is only the first step, and now the real work has to start. It is necessary to invest time and effort in developing trust and open exchange of information. This involves building up contacts and joint team working at several levels of the partner organizations. I would identify five discrete levels that need to be covered as:

- Routine Operational
- Tactical Planning
- Financial and Performance Monitoring
- Strategic Development
- Corporate Bonding.

Routine Operational is the day to day exchange of the vast amount of information that is needed to carry out the work, and to show that it is being carried out. In this, of course, the IT system plays a vital part. Equally important is the human contact that handles routine and unpredicted events, gives the necessary speed of response and ensures the level of service. This human contact has to be not only at Manager, Supervisor and Traffic Controller level, but also between drivers and stockhandlers with sales floor or production staff, so that they begin to identify with and take a pride in the overall effectiveness of the combined operation.

Tactical Planning is the work carried out by joint teams to ensure that special requirements, such as Christmas peaks or factory shut-downs, are effectively covered. These exercises, usually involving many meetings spread over several months, offer a real opportunity for team-work and exchange of know-how and should contribute greatly to the build up of confidence and trust.

Financial and Performance Monitoring involves periodic meetings of the middle management teams of the partners. Inevitably there is an adversarial element in these sessions, haggling over budgets, attacking and justifying variances, but with time even here the atmosphere can become more open and constructive as the importance of the common aim begins to emerge.

As the partnership develops, Senior Management should begin to step back from detailed involvement in the operational, tactical and monitoring activities, and begin to work together on the strategic development of the supply chain. This will require considerable input and commitment from both sides for which no immediate return can be expected, but relies on the trust that if and when there is a dividend, it will be fairly apportioned between the parties.

Finally, at Top Management levels there needs to be regular, if infrequent, meetings. While these may not discuss major issues of substance, they are a form of Corporate Bonding which allows each to understand better the other partner's culture and drives. It also serves as an important demonstration to their respective organizations of the commitment to the partnership.

The first three of these levels will normally begin by 'working to contract', but it is these levels of co-operation that will generate the 1000 tiny advances that will together give major improvements in performance and speeding up responsiveness of the existing supply chain. Significant contacts at the Strategic Development and Corporate Bonding levels will only begin to take place once the partnership has developed to a more mature level, largely through the demonstrated successes of lower levels.

Mention was made earlier of how a common information system is the spinal chord of any integrated logistics partnership. But all such a system does is move information. At the end of the day it is the movement of product that counts, and product is still largely moved by people. To be successful, those contractors working in the critical/complicated segment of the contract distribution market, must attach the highest importance to the training, development and motivation of all their workforce. It is important that stockhandlers and drivers as well as supervisors and managers feel that they work for the client as well as for the contractor and that they are all seeking ways of improving the overall effectiveness of the operation. There is a cultural dichotomy in any modern distribution operation – 99% of activities are controlled by strict disciplines and procedures, in fact usually driven by a highly impersonal computer terminal. How else can one ensure 99+% picking accuracy and 97% on time deliveries? Yet the same staff have to be empowered and have to have the initiative to take action to sort out the 1% of activities which do not conform. All that this requires is good management and leadership at all levels, but that in turn requires a massive investment in training and real commitment to creating the right culture at all levels.

8   TWO DILEMMAS

In thinking about the theoretical concept of partnerships there seem to be two dilemmas for which totally satisfactory answers have not yet been found. The first of these I call the Pied Piper syndrome. When the town of Hamlyn was infested with rats, the Mayor and Corporation were quite happy to offer forty thousand ducats to whoever could rid them of the menace. However, when they saw how apparently easy a task this was for the Piper, they reneged on their side of the bargain. Parallels can be drawn with some developments in the contract distribution industry. The lesson here is perhaps that the clients should be prepared to pay more initially for the major change and innovation which contracting out can bring, but equally the contractor should not realistically expect to be able to go on charging premium rates once the operation has settled down to a more or less steady state.

The second dilemma is the question of exclusivity. If a client is persuaded to enter a partnership with a distribution contractor in the belief that the latter can make a significant contribution to the client's competitive edge, then the client will not be happy to think that the contractor may go off and do a similar job for his competitor. The advantage for which he has paid is immediately cancelled out. If any contractor is really serious about offering its clients competitive advantage, then logic suggests that it needs to impose upon itself the discipline of only working for one major client in each market sector. On the other hand, those big clients who generally like to spread their risks by employing several contractors, possibly while retaining some operations in-house, can themselves undermine their contractors' competitive edge if they insist on spreading best practice from the good contractor to the less good. As with any investment, diversity reduces risk, but in the process sacrifices the chance of the biggest wins.

9   CONCLUSION

In reality, of course, partnerships do not always follow the smooth path I have described. Recession, conflict of interest, mistrust, major structural changes, etc. all put stresses and strains on the relationships. However, the fact that partnerships do survive such traumas and may even be strengthened by them leads me to believe that contracting out on a partnership basis is the bet way forward for many companies whose logistics are both complex and critical to their competitive edge.

8 Distribution Trends in the Food Manufacturing and Processing Industries

M. Browne[+] and N. Greenslade[*]
[+]University of Westminster
[*]Exel Logistics

1 INTRODUCTION

Few, within the logistics industry, would deny that the major food retailers have been instrumental in effecting change within the Grocery Supply Chain. Their ongoing strategies to drive down costs, both internally and with suppliers, have generally been respected if not always welcomed.

This emphasis on cost reduction as part of a continuous search for competitive advantage has to be set against a background of unprecedented growth in the number of new retail outlets. In recent years as many as 75 new superstores have been opened each year in Britain and the country now supports around 900 superstores.

The media attention paid to the high profile food retail market has however overshadowed the advances and efficiencies made by many food manufacturers to not just reactively meet customer demands, but recognize distribution trends and anticipate the impact on their own production and distribution disciplines. It is acknowledged however that manufacturers acceptance of some retailer strategies have only been as a result of competitive threat and at the expense of profit margins.

A recent report by Andersen Consulting confirmed that although logistics had emerged as a business function in its own right only half of the 261 companies surveyed knew their total logistics costs. In a mature industry understanding and controlling your costs and anticipating trends is essential in order to remain competitive.

2 IMPORTANCE OF DISTRIBUTION FOR FOOD MANUFACTURERS

2.1 Costs

In recent years however food manufacturers have been strikingly successful at reducing their logistics costs. The 1990/91 ILDM Survey of Distribution Costs suggested that food manufacturers distribution costs amounted to 7.29% of turnover. The following Survey covering 1991/92 showed that this figure had fallen to 4.53%. However, this average figure masks some important differences according to the size of company concerned (Table 8.1).

TABLE 8.1. Distribution Costs as a Percentage of Turnover for Food Drink and Tobacco Manufacturers.

Co. size	Storage	Inventory	Transport	Admin/ Packaging	Total
Small	3.4	1.1	4.6	0.2	9.3
Medium	2.1	0.4	2.1	0.1	4.7
Large	1.2	0.7	2.2	0.1	4.1
Average	1.5	0.7	2.3	0.1	4.5

Source: ILDM Survey of Distribution Costs 1991/92.

The analysis would suggest that smaller companies have not yet achieved the economies of scale open to larger companies but it may also reflect the lower priority placed on controlling logistics costs over production improvements for example.

2.2   Operating margins

Smaller companies have also fared worse in terms of overall operating profit margins. From an industry average of 7.1% for three years, in the past year there has been a decline from 6.7% to 6.3%, the lowest level since 1987.

Particularly affected were companies with an annual turnover of less than £100 million while 40% of larger companies actually increased their margins (OC&C Strategy Consultants/Coopers & Lybrand, 1993).

3   KEY TRENDS IN DISTRIBUTION

Changes in food retailing strategies inevitably influence food manufacturing. Indeed the inter-dependence of the manufacturers and retailers and the need to try to work more closely together, has been an important thread in many of the logistics developments that have been noted during the 1980s and early 1990s. Among the most significant trends influencing the logistics strategies of food manufacturers have been: the growing integration of the supply chain, a shift in the ownership and control of stock, changes in consignment size and delivery frequency and the development of, and growing reliance on, more powerful information technology and communications systems. Each of these influences is discussed in greater detail below.

3.1   Integration of the supply chain

Writing about changes in the marketing channel for retailers, Dawson (1993) noted that:

> '... the structure of the marketing channel has also changed from being essentially a pipeline along which goods flowed to being a set of relationships amongst the various organizations involved.'

Integrating these relationships has become a key management task. Although functions such as product design, branding, promotion, selling and so on remain largely the same the way they are performed has changed and so has the structure within which they are performed. As Dawson goes on to say:

> '... the activities have become more integrated and co-ordinated rather than being undertaken in several different organizations (manufacturers, wholesales, retailers) each of which has its individual combination of activities.'

However, it is also clear that the nature of the relationship between retailer and manufacturer varies widely depending on industry concentration and the balance of power.

Andersen Consulting (1992) have suggested that there has been a rise in partnerships between retailers and manufacturers initially concerned with operational improvements in packaging and delivery patterns. Manufacturers have however also sought and developed partnerships with other manufacturers to consider the scope of developing common transport arrangements to minimize cost.

3.2   Control of stock

During the 1980s all industries have tried, in many cases with considerable success, to drive stock out of the supply chain. In some instances it is probable that stock has not really been driven out of the supply chain, but simply pushed along it until the weakest player in the chain had no choice but to accept the stockholding responsibility. However, even if this less than ideal solution were true, there would still be some cost advantages since stock held upstream in the supply chain will typically have a lower inventory holding cost as it has been subject to fewer processes which add value. Much more significant gains have been made when all members in a supply chain have used better management techniques and the advances in the transport and distribution industry to reduce stock and successfully drive down inventory holding costs; usually without sacrificing customer service and in some instances improving it.

A trend observed in the grocery sector has been the shift to greater control and ownership of stock by manufacturers. Increasingly, it appears that retailers will expect stock to be held at manufacturers sites. Andersen Consulting (1992) reported that 46% of retailers and distributors responding to a food industry survey believed that stock will be held at the manufacturer.

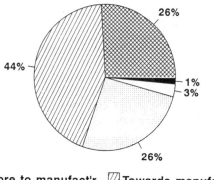

🔲 Far more to manufact'r   🔲 Towards manufacturer
🔲 Remain about the same   🔲 Towards retailer
■ Far more to retailer

FIGURE 8.1. How do you think the ownership of stock will shift over the next two years?
Source: Food Industry News Survey, 1993.

Looking at some more recent data specifically from the food manufacturing sector, the shift to greater responsibility being with the manufacturer becomes even more strongly evident (see Figure 8.1).

A key issue revolves around the reason for this shift in responsibility. The following factors have been important in driving the changes in stock holding strategy among the partners in the food supply chain.

- **Retailer power:** Broadly speaking retail power is becoming more concentrated and this has been much in evidence in the grocery sector. Five firm concentration ratios (i.e. the share of the market held by the five largest companies in a sector) have increased, both in the UK and in other European countries. It has been suggested that 80% of UK food volume is through both the major multiples and other recognized multiples. Increasing importance in own brands or private label products has also influenced the relative power of retailers and manufacturers (about 60% of Tesco and Sainsbury sales come from private-label products). The power of the retailers has meant that they have simply been able to insist that their suppliers control the stock and retain ownership.
- **Stock reduction strategies:** Growing competition at a national level has encouraged both retailers and manufacturers to look at ways to reduce costs in order to improve margins. Stock holding costs can represent a significant proportion of total logistics costs and can have a major impact on profitability. The level of stock reduction will depend on the confidence retailers have in manufacturers meeting their demanding service levels in

terms of responsiveness, accuracy and reliability. The lead time from order to supply will become shorter and there again must be considerable trust between manufacturer and retailer. Promotions and bulk buys can also be incompatible with a minimum stock strategy. While this may have seemed a rather limited barrier to development in the late 1980s and early 1990s the growth of discount operations may induce retailers to reconsider the strategic importance of bulk buys as they seek new ways to ensure continued customer loyalty.

Short shelf life products are becoming more important in food retailing and in many cases this has reduced the amount of product held at any one time in the supply chain. Food manufacturers are likely to assume growing responsibility for stock control in this sector.

- **Manufacturing strategies:** Some large manufacturers are seeking scale benefits by producing a smaller range of products at each of their national production plants (sometimes referred to as a focus factory approach). One result of this has been the desire to postpone the stage at which a product is dedicated to a specific market. In order to achieve this, packaging, for example, will be postponed as long as possible or packaging that can be used in several countries may be adopted. Manufacturers of canned products may label products on a small operation within the warehouse just before despatch. This may especially apply to export operations where the product is standard but the label unique in which case every production run may include an amount of 'bright' or unlabelled product. These developments have been less pronounced in the food sector than in other areas of fast moving consumer goods such as household products and toiletries, they nonetheless illustrate how manufacturing strategies can lead to the supplier taking greater responsibility for stock in the supply chain.

Just-in-time trends and the growth of lean production strategies has received widespread attention in the automotive industry. However, there is also evidence that food manufacturers are using new manufacturing techniques enhancing the opportunity for small batch production, becoming more responsive to fluctuating customer demands and reducing stock in the supply chain.

If trends in Japan are copied elsewhere, then the importance of rapid response to changes in consumer demand will grow. For example, in the soft drinks industry within Japan, more than 700 new products and brands are marketed each year, but about 90% of them disappear after only one year in the market (Ohbora *et al.*, 1993).

3.3 Consignment size and delivery frequency

In a more responsive environment with continued emphasis on stock reduction there will be pressure on food manufacturers to reduce consignment size. However, a reduction in consignment size, typically, increases the unit cost of delivery.

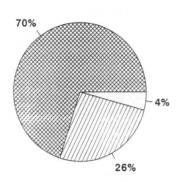

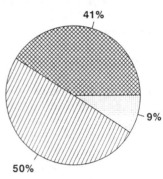

▩ Increase ▨ Remain the same ☐ Decrease

FIGURE 8.2. How do you believe the frequency of consignments delivered will change over the next two years?
Source: Food Industry News Survey, 1993.

The frequency of deliveries will also have to increase given the same total volume moving through the supply chain. Clearly this makes financial sense if savings in inventory holding and storage costs are sufficient to outweigh any additional spending on transport.

A recent survey of food manufacturers (Food Industry News Survey, 1993) revealed that many expected a strengthening of the move towards more frequent deliveries (Figure 8.2).

The need to deliver smaller consignments more frequently will encourage manufacturers to share information to allow stock consolidation of several suppliers prior to delivery to an RDC. Again this development is born out of a need to enhance vehicle utilization and minimize delivery costs. The trust implications for manufacturers in a competitive market may necessitate a third party logistics contractor acting as an 'honest broker'.

Delivery frequency will be tied to shorter lead times between order and delivery. The principle of 'cross-docking' or 'over-the-bank' operations already adopted for perishable products will expand with pre-assembled orders not an impossible future requirement.

3.4   Information technology and communications systems

Sharing information between manufacturer and retailer by electronic means has been a central feature of developments in the food supply chain. Concrete examples of the benefits have been well documented. For instance following the implementation of a system to integrate scanned data and computer facilities with suppliers' systems, Sainsbury reported a decrease in order cycle time from two weeks to one (Council of Logistics Management, 1993).

Over the next few years there will be further rapid growth in the importance of electronic systems in the food supply chain. Continued decline in hardware costs will encourage small and medium sized retailers to install systems that have been seen only in very large organizations (Brown, 1992). The spread of computerized systems and scanners at the checkout of, say, corner shops will enable many more retailers to gain a fuller understanding of their stock and stock movements. This may then further increase the demand for smaller, faster stock replenishment at numerous, wide spread small retail outlets. This will again stretch delivery networks across the country.

Changes in the provision and use of electronic information systems has many implications for food manufacturers. Recent comments by Angus Clark of Sainsbury suggested that as larger retailers continue to develop their systems there may be a move to maximize the efficiency of primary distribution from the source of manufacture to the retail depot. The implication being that it is in this area of primary transport from manufacturer to RDC that there is greatest immediate scope for further efficiency. Standardization of pack bar codes already exists, but the opportunities to track consignments in pallet or cage quantities across different service providers will mean the development of new systems throughout the supply chain. Some retailers already track cages through the operation providing 'real time' location information and activate production of vehicle manifests and way-bills.

However, as more smaller and medium sized retailers increasingly adopt electronic information systems there will be a need for manufacturers to integrate their systems with more customers. Indeed manufacturers themselves expect the degree of integration to increase (Figure 8.3). A retailing concept from the United States of America, Efficient Consumer Response (ECR) relies on partnerships being formed between Manufacturers and Retailers working closely together to optimize stock levels. Accurate and timely information is therefore a key issue to support effective marketing, production and logistics decisions.

4   CONCLUSIONS

Among the developments described there are four trends worth re-emphasizing.

- Far less storage space will be needed (or provided) at the retail outlet and in the same way retailers RDCs will become increasingly a focus for transhipment and rapid product turnaround.
- Quick Response will place more pressure on the transport operation.
- As stock is removed from the system and is pushed back upstream in the supply chain there is increased potential for stock-out problems unless demand is accurately forecasted and information communicated swiftly.

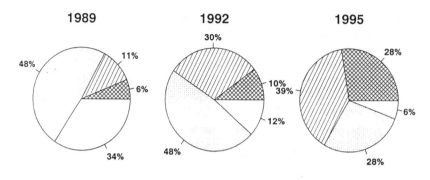

⊠Very high ⧄High ▢Some ☐Totally separate

FIGURE 8.3. Degree of integration with customers.
Source: Food Industry News Survey, 1993.

- There will be an ever growing need to integrate the information flows as well as the operations of the partners in the supply chain.

The need to drive down costs and retain competitive advantage will ensure that quick response remains a fundamental feature of the logistics supply chain. With a changing market place, the need to maintain strong working relationships and openness in information sharing will be essential.

REFERENCES

Andersen Consulting (1992) Grocery Distribution in the 90's, A research study conducted for The Coca-Cola Retailing Research Group – Europe.

Clark, A. (1993) Leaders of the pack, *International Freighting Weekly*, 29 March.

Dawson, J. A. (1993) Issues and trends in European retailing and their impact on logistics, Proceedings of the ILDM National Conference, June.

Food Industry News Survey (1993) Conducted by the University of Westminster.

Institute of Logistics and Distribution Management (1992) Survey of Distribution Costs, ILDM:Corby.

Ohbaro, T., Parsons, A. and Riesenbeck, H. (1992) Alternative routes to global marketing, *The McKinsey Quarterly*, 3, pp. 52-74.

OC&C Strategy Consultants/Coopers & Lybrand 1993.

9 Pan-European Business Systems: Results of Survey

T. Evans
Tony Evans Associates

1 INTRODUCTION

The key issue in today's business environment is the reconciliation of the often incompatible pressures of cost control, responsiveness to competition and the drive to develop new markets.

Faced with such pressures and the establishment of the single European market, change and rationalization seem inevitable. Mergers, take-overs and even business failures are set to occur as competitive pressures mount. In consequence, more truly pan-European organizations, with operating units spread across the continent, are now starting to emerge.

Europe's leading organizations agree that competitive advantage is gained in today's turbulent business environment by improving the quality of both customer service and products. And, as most organizations accept, the only way to meet the challenges of improving quality, while maintaining cost control and competitiveness, is through flexibility. Such flexibility, both in working practices and market orientation, requires a timely, effective and accurate flow of information.

One obvious way to ensure an effective and efficient flow of information is through the implementation of common systems. Common procedures, common terminology and common reporting mechanisms are prerequisites of effective, rapid decision-making in successful pan-European organizations. The arguments for adopting common systems are strong and self-evident; they enable quick and effective decisions; the information technology needed to drive these systems can be matched to organizational requirements and the move to a single European market has further removed the barriers to running an effective pan-European operation.

Yet despite the clear arguments in favour of common systems, a recently commissioned survey of over 150 CEO's and CFO's from the Financial Times Top 500 European Companies, has unearthed a reluctance on the part of their organizations to adopt common systems outside the financial systems and performance reporting areas and has further explored the reasons behind that reluctance. The survey reveals that the main obstacle to common systems is not a fear of technology, but the attitudes of subsidiary companies. Managers of subsidiaries, viewing information as their power base, feel their autonomy would be threatened and that they would be subjected to ever tighter control as a consequence of implementing common operational and planning systems.

The survey has shown that whilst considerable progress has been made with the adoption of common financial systems and common systems for performance reporting, most companies are still at the strategy and planning stages in other application areas. The lack of progress in areas such as manufacturing and supply chain management generally, has been explored and the main factors have been identified.

Responses to the survey were received from companies across a wide range of manufacturing and service sectors, including automotives, chemicals, pharmaceuticals, food, electronics and transport. Countries covered by the survey comprised Belgium, Denmark, France, Germany, Netherlands, Italy, Spain, Switzerland, Sweden and the UK. Overall, there was universal agreement regarding general business pressures in the 1990s, however there were interesting contrasts of attitude concerning the implementation of common or pan-European systems. For example, Swedish and Dutch subsidiaries are generally more receptive to the adoption of common systems that other nationalities.

Generally, it was agreed that the importance of IT to support the introduction of new streamlined pan-European businesses is creating a trend to involve IT Directors and Managers at a strategic level.

The survey suggests that systems which are centrally-imposed are not used co-operatively. It would appear, therefore, that the best way of overcoming resistance to common systems is to ensure that the benefits are explained clearly in advance; that training and support is provided to raise staff confidence and that an ongoing communication programme maintains momentum before, during and after any systems are put in place.

2   CURRENT BUSINESS PRESSURES

European businesses are involved in the difficult balancing act of fighting off competition in previously secure markets, while maintaining tight control on costs. Easing of trade restrictions and improving global communications mean that businesses now face the competition not only of other Europeans, but also of predators from further afield. Fighting harder to secure local trade, let alone extending into other markets, significantly raises costs. Cost control has therefore become an issue of primary concern among the companies surveyed.

Both cost control and increasing competition are, however, issues which focus companies' attention internally at the expense of external issues such as customer needs. Yet executives are aware that improving standards and increasing responsiveness to customers product and service requirements are issues equally deserving of attention. In difficult trading conditions, it is often a buyer's markets with buyers dictating the standards for higher levels of service or specifications and lower prices. Suppliers must then try to satisfy the market more fully than their competitors without incurring too great a cost burden (see Figure 9.1).

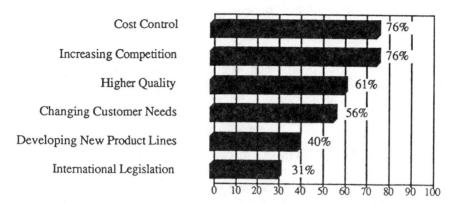

FIGURE 9.1. Business pressure (% respondents viewing the issue as highly important).

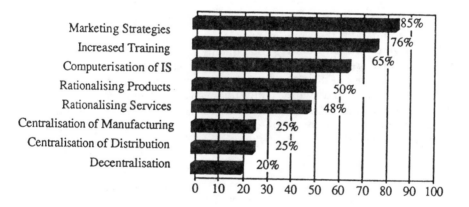

FIGURE 9.2. How companies are responding to market pressures.

3 COMPANIES' RESPONSE TO MARKET PRESSURES

The encouraging message from this survey is that European businesses are trying to take positive steps to ensure their long term success. The most common response to the threat of increased competition is to look for appropriate marketing strategies. Such strategies are designed to increase sales and to grow market share, the benefits of which can then be exploited when trading conditions improve. In order to trade more effectively, sales and marketing practices and also products need to change and to enable this to happen the issue of staff training must be addressed. Companies are reacting to the skill shortages which arise from changes in commercial activity by retraining existing staff, thus making the best use of existing resources.

Having chosen to redirect their efforts towards effective marketing and selling, supported by appropriately skilled staff, the next significant factor is recognizing the important part played by information in co-ordinating these activities. This is particularly true in multi-national companies where control of a highly dispersed operation can be achieved through the imaginative use of technology (see Figure 9.2).

4   BENEFITS TO BE GAINED FROM COMMON SYSTEMS

Organizations report a wide range of benefits which result from common methods or systems. In many cases, these benefits are relatively recent as the changes are still in progress. However, those companies who have already succeeded in implementing common systems are enthusiastic about the results.

Reported benefits include:

- **The streamlining effect of common systems:** The practicality of handling common data in the same manner.
- **Positive effects on quality:** Total quality management – as standards and procedures are clearly visible in common systems.
- **Ease of document distribution:** Ease of communication, whereby documentation is exchanged swiftly with little need for modification.
- **The application of shared standards for improved quality:** Maintenance of the highest quality of products in all markets by the application of approved standards.

The general impression gained from all interviews conducted during this survey is that a company can adopt a consistent approach both to its internal activities and to customers and suppliers. Typical benefits resulting from this approach include improved performance, better control and the elimination of errors or confusion caused by inconsistent data.

5   BARRIERS TO IMPLEMENTING COMMON SYSTEMS

The barriers to implementing common systems across an international organization are found in a number of quarters. Most commonly, however, problems arise from national or cultural differences existing between companies in different locations or countries. Some 85% of respondents admitted that a tension exists between pursuing common standards and practices and maintaining the perceived independence of offices and/or companies within the organization.

European unity is therefore not easy to achieve and several large organizations

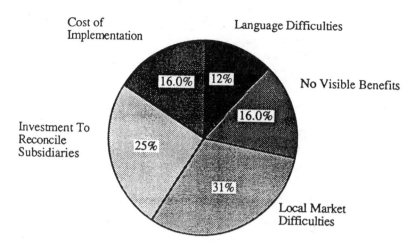

FIGURE 9.3. Barriers to implementing common systems.

are clearly not actively pursuing it. Their reasons principally echo the above mess-age in that they wish to avoid difficulties in local markets and not invest large amounts of time, money and effort in reconciling differences between subsidiary companies (see Figure 9.3).

These responses imply that considerable inertia exists in many companies who would prefer to weather the storm in their present condition and defer systems development until conditions are more favourable. The counter argument to this stance is that preparation at this stage allows companies to benefit fully from any improvement in trading, without diverting resources in the 'catching up' process.

6   CURRENT PROGRESS IN IMPLEMENTING COMMON SYSTEMS

A common or pan-European system is recognized, for the purposes of this survey, to be an IT system operating in a number of locations and in different countries which uses the same basic software facilities (see Figure 9.4). What is implied is a planned process of implementation, central co-ordination, and, usually, a high-level relationship with software authors or vendors in order to provide a consistent supply of products to all users, irrespective of hardware platform.

Only 8% of respondents have fully completed such an implementation pro-cess, whilst 74% of companies report that they are actively planning or imple-menting such systems. Responses to this survey suggest, that while the benefits of common systems are widely accepted at the strategic planning level, these bene-fits are not always successfully communicated throughout the rest of the company.

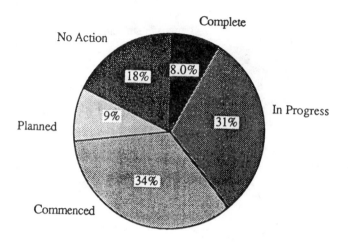

FIGURE 9.4. Implementing status of common systems.

This failure may partially account for the gap between the number of companies who plan systems and those who successfully complete the implementation process.

7   APPLICATIONS CURRENTLY IMPLEMENTED

This survey indicates that the majority of common system implementations involve financial or management information systems in some form. This may be explained in a number of ways.

Financial systems are often the first modules to be implemented in manufacturing companies under a more widespread plan. Ownership or sponsorship of such systems also tends to be concentrated in the financial hierarchy, where benefits can be quantified and more readily balanced against costs.

It can also be argued that financial disciplines are the first to receive systems attention as they are already more standardized internationally than, for example, manufacturing or warehousing systems. The latter are subject to more variations arising from custom and practice at a shop level and it is perhaps harder to convince managers, who may feel threatened by the prospect of harmonization and working more closely with other companies or nationalities, of the benefits of a common system. Management information is a logical application to implement across multiple operations as it allows the business managers to develop a fuller picture of the state of the organization.

The next, and perhaps the most testing initiative, according to this survey, is to move on from the relatively easy first steps of finance and MIS. Supply chain

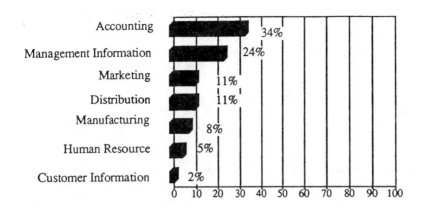

FIGURE 9.5. Implementing progress of common systems applications (in companies who have COMPLETED implementation).

management systems offer potentially far greater rewards, but correspondingly far greater challenges, particularly in companies which are planning to introduce new pan European logistics networks and facilities.

8   CONTRADICTIONS PINPOINTED BY THE SURVEY

Drawing a link between several of the findings discussed here, a clear disparity arises between business theory and systems practice among our responding companies.

Some 85% of respondents claimed to be reacting to the pressures of international competition and cost control with enhanced marketing and sales strategies. In addition 65% are developing their IT systems and the survey shows that further large-scale investment in IT is planned.

The contradiction, however, lies in the fact that the bulk of systems investments is being made outside of the marketing area. The survey has shown that only just one in ten companies have implemented common marketing systems, and one in fifty in the customer information area (see Figure 9.5).

Lack of progress in these areas is probably explained by the fact that marketing and customer information systems depend on information gleaned form outside the company and a high degree of co-ordination and co-operation with external customers and suppliers, and are therefore a far more complex proposition than an internally focused financial system.

9   CONCLUSIONS

This survey shows that there is now a wide acceptance at board level that common practices and systems are a legitimate goal for multi-national companies under current business conditions. Common IT systems have delivered real benefits to those who have taken this step.

If, as the survey indicates, financial systems are the logical and relatively easy first step in the implementation process, then the next task is to set the agenda for, and prepare to face, the difficulties associated with the second stage of this process.

The second stage will almost certainly be linked with the operational side of the organization. It will probably cover sales and marketing support, manufacturing and distribution, in new totally integrated pan-European supply chain management systems, which have been designed to support a rationalized and process oriented pan European business structure. The implementation of such systems will allow companies to reduce inventing holding costs whilst offering a level of total customer service which was unavailable under the fragmented practices of the past.

Throughout this process, success will owe much to the level and quality of planning which is applied to these projects. Co-ordination of activities on such a wide scale is a challenge to the management teams of these organizations. This is doubly so when the effects of organizational and system changes reach the furthest extremes of factory floors, warehouses, and individual salesmen.

And it is at just that individual and factory level according to the survey, that the benefits of common systems, while ultimately greater than elsewhere, are less immediately obvious or personal. Sensitive handling of the change processes in these areas will therefore contribute dramatically to the success of these projects, both locally and centrally.

10 Supply Chain Planning Under Uncertainty

C. Lucas*, G. Mitra* and S. A. Mirhassani+
*Brunel University and Unicom Consulants
+Brunel University

1 BACKGROUND

1.1 An analysis of issues

Many organizations are faced with a constantly changing, uncertain economic environment and progressively shorter product life cycles. In addition complex corporate joint venture alternatives need to be evaluated and requests for increasingly stiffer service levels have to be taken into consideration.

In order to respond to these challenges managers and planners in a variety of companies are increasingly turning to integrated logistics planning. Taking into account the supply chain activities of manufacturing, processing and distribution, it is easily seen that both long term strategic planning and shorter term tactical scheduling decisions closely interact in finding effective and robust solutions (see Figure 10.1).

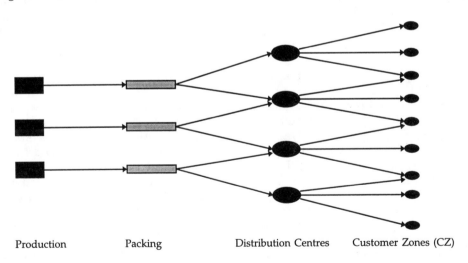

Production   Packing   Distribution Centres   Customer Zones (CZ)

FIGURE 10.1.

Four key aspects of this problem are identified as:

- time;
- uncertainty;
- total cost; and
- customer service level.

These factors influence the planner's choice and managerial decisions. The interaction of current plant capacity (current assets) as planned today and future demand taking place as the future unfolds, makes 'time' a critical factor and the randomness of the demand and plant behaviour makes 'uncertainty' the second critical factor. Costs must be taken into account in 'totality', that is fixed and variable costs must be considered and finally the sensitivity of 'service level' to alternative decisions has to be analysed. Most organizations are turning to model based decision making to analyse their options and choices. The models, must encompass these aspects in some quantifiable form. An integrated model impacts on other aspects of organizational operations which can no longer be taken in isolation. Thus, in-house or external vendor production, local or global outsourcing for a multinational, cross border trade: the impact of exchange rate and respective tax regime on profitability are related corporate problems. These complex options can be captured and analysed only by a model based approach.

1.2   Integrated approach

From the analysis of the issues above it is easily seen why a class of integrated planning models are of growing interest to senior management in manufacturing firms. To start with, we have to capture a closer coupling between strategic planning and tactical scheduling decisions which follow as an inevitable consequence of a shorter product life cycle. A global multinational player would ideally like to take business decisions which span sourcing, manufacturing, packing/assembly, distribution and product pricing. Thus a company with multiple production plants, and multiple markets may seek to allocate demand quantities to different plants over the next month, next quarter or next year time horizon. Its objective is to minimize the sum of manufacturing and distribution costs associated with satisfying customer demands. Alternatively, if the product mix is allowed to vary, the company may seek to maximize net revenue. Each plant is described as a particular entity with respect to: its direct, indirect and overhead costs; its resources including machine capacities, labour and raw materials; and its recipes for producing products from raw materials and other resources. Direct and indirect costs may be characterized by economies or diseconomies of scale, and by fixed costs. The model also captures the flow of products from the plants through warehouses to the markets. By combining the latest developments in OR decision models, analytic databases or data warehouses (see below) and powerful desktop computers it is possible to construct integrated decision support systems (Shapiro *et al.*, 1992; Shapiro, 1993). Such a system can be a collection of interrelated models designed to answer questions such as:

- How can the total distribution cost be minimized?
- What is the trade-off between response time and cost?
- How many warehouses should we have? Where?
- How should each warehouse be supplied?
- Which warehouse should serve which customer?
- What is the added cost of sole sourcing each customer?
- Which products should come from which sources?
- How much inventory should be carried? Where?

It is of course the goal of a managerial team of decision makers and analysts to investigate and answer such questions as clearly and comprehensively as possible so that robust and optimum asset allocation and operational logistic decisions can be made for the supply chain management. The availability of a class of such models can also support the managerial decision makers in other problems which might not have been in the original scope of their investigation. Typical examples are:

- designing the sourcing of a new product;
- impact of exchange rate and taxation policy changes;
- analysing mergers and acquisitions; and
- coordinating new marketing strategy with logistic support.

1.3 Analytic decision databases

Good decision making relies upon readily available data (or facts) and on good quality analytic information. It has now been accepted by information systems specialists that operational or transactional databases which support the organization are not necessarily suitable for decision making. There is a strong move towards creating analytic databases or data warehouse for the organization (Codd *et al.*, 1993; Codd and Codd, 1994). Shapiro calls it the decision database as it supports decision making models. In some sense the understanding and creation of the analytic database through the information value chain (see Figure 10.2) is as important as the models outlined earlier.

Typically, a transactional database has information about plants production rates, costs, warehouses, and order quantities. An external database could be map information, routes, customer location by postcode, or hauliers for outsourced distribution. Analysis of data items lead to forecast of demands, plant performance efficiency, breakdown rates. External analytic information could be competitor intelligence, industry sector performance indices, economic forecasts, or consumer behaviour trends.

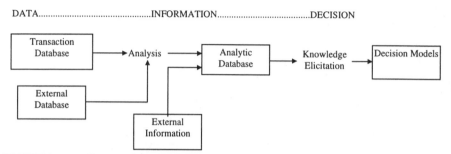

FIGURE 10.2. Data-Information-Decision value chain.

Thus creating and understanding the analytic database and the progression through the data, analysis, information and decision model value chain is an important aspect of the organizational decision support system.

1.4   The role of a stochastic programming (SP) Model

Planning and management of the supply chain are underpinned by 'time' phased decisions taken in the face of 'uncertainty'. In the planning network, uncertainty propagates from sourcing to customer demand, see Figure 10.3. The typical sources of uncertainties are late delivery of raw material, machine breakdown, or revision of the order quantities (or order cancellations). In strategic models this calls for overcapacity of plants and in logistic models this means increased inventories; these are simple insurances against uncertainties. Whereas linear and integer programming are now well established and deliver practical solutions in deterministic situations they are clearly not suitable for this context. Although LP sensitivity analysis sheds some light on the behaviour of the solution, this approach is still inadequate to analyse the distribution or the robustness of the optimum solution. Recently, there has been considerable development (Mulvey *et al.*, 1995; Birge; Dempster and Thompson, 1996; Lucas *et al.*, 1995; King, 1994) in the modelling and computational solution of stochastic programming (SP) problems. For this class of problems SP models are well suited to capture the uncertainties in the supply chain network but the following real issues still remain:

- how much information about the probability distribution is required to arrive at a robust (well hedged) optimum decision?
- what does the term 'optimum decision' mean in this context?
- how should one construct scenarios and analyse the scenario based decisions?
- how can the solution of the SP be computed within the available computer processing power?

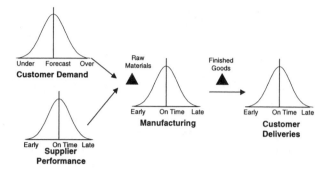

FIGURE 10.3. Uncertainty in the supply chain.

2 MODELLING AND SYSTEM DEVELOPMENT ISSUES

2.1 Creating and elicitating a generic decision model

The supply chain management problems have been cast in the form of deterministic mathematical optimization models and many real instances have been computationally well solved. But constructing a generic model which closely couples the strategic planning and tactical logistic decisions as well as capturing the 'time' phasing and the uncertainty elements of the sourcing and the supply chain problem remains a challenging task. In particular for the model to be usable in a managerial setting one must explain the abstract structure of the model to the decision makers and engage them in a participative model development process. By involving the decision making team in this way we can not only elicit their domain knowledge but can also gain their confidence and commitment to the computer based modelling support tool.

2.2 Creating the analytic database

Prior to the elicitation of the domain knowledge the development of the symbolic decision model has to be closely coupled to the available information. Again the decision makers have to be engaged in a participative role of compiling, structuring and analysing the facts, data items and information that is available. Making use of suitable OLAP (including multidimensional analysis) tools and middleware tools (see Bloor Research Report; Sach *et al.*, 1996) the transaction oriented production data bases can be processed and a suitable analytic database (data warehouse) can be created. Analysis of the data, choice and use of tools to create meta data, juxtaposition and analysis of the model based plans and decisions (solutions) in relation to the original data call for considerable investigative and research work.

3   A CAPACITY PLANNING MODEL

3.1   A supply chain model as a SP application

Having obtained the analytic database, it is then necessary to specify the underlying deterministic optimization problem. In this section, we detail our experience with a real life application in supply chain planning. The model developed represents the entire manufacturing chain, from the acquisition of raw material to the delivery of final products. The following describes the deterministic model and later we address the issue of incorporating uncertainty. In order to capture the interactions that exist among different stages of the manufacturing process (Davis, 1993), at each time period $t \in T$ the chain is simplified to four stages relating to production, packing, distribution and customer zones.

We let *PR*, *PC*, *DC* and *CZ* denote the sets of production plants, packing plants, distribution centres and customer zones. Production and packing plants consist of several lines, with distribution responsible for deliveries between sites and customer zones. Different types of a product can be produced and distributed.

The result is a multi-period multistage model which captures the main strategic characteristics of the problem as, site location decisions, choices of production and packing lines and capacity increment or decrement policies.

In order to formally state the decision problem we define the sets, indices and variables used in the mathematical formulation as follows.

> i denotes the sites at which production and packing plants are located;
> $l \in L_i$ denotes the type of line where L_i is the set of line types at site i;
> M_i is the maximum number of lines allowed at site i;
> $k \in DC$ denotes distribution centres;
> $h \in CZ$ denotes customer zones;
> $p \in P$ denotes the product types;
> $t \in T$ denotes time periods;

The decision variables at each time period $t \in T$ are the following:

> $z_{ti} \in \{0,1\}$ indicates if site i is open;
> $r_{til} \in \{0,1\}$ indicates if a line of type l is open at site i;
> $C_{til} \in Z^+$ represents the number of lines of type l open at site i;
> $x_{til} \in Z^+$ is the number of lines of type l opened at site i;
> $\theta_{til} \in Z^+$ is the number of lines of type l removed before the end of their life time at site i;
> $Q_{tkp} \in Z^+$ represents the ordering quantity of product p at the distribution centre k;
> $F_{tkhp} \in Z^+$ is the number of product units of type p shipped from k to h;
> $q_{tilp} \in Z^+$ is the number of product units of type p packed by a line of type l at site i;

$\mu_{thp} \in Z^+$ represents the shortage of product p at customer zone h.

In many cases it has been shown that is important to take into account uncertainty in the inventory management of manufacturing systems (Cohen *et al.*, 1990; Schoemaker, 1988). In particular, distribution centres must be managed so that they achieve minimum inventory costs while meeting expected customer demand.

Let D_{thp} be the customer demand in time period t for product p at customer zone h. Then the following constraint must be satisfied:

$$\mu_{thp} \geq D_{thp} - \sum_k F_{tkhp} \qquad \forall\ t,h,p \qquad (1)$$

where the shortage of a product μ_{thp} is minimized by including μ_{thp} in the objective function.

Distribution centres must make a decision about order quantities Q_{tkp} such that:

$$\sum_h F_{tkhp} = Q_{tkp} \qquad \forall\ t,h,p \qquad (2)$$

Note that constraints (1)–(2) represent the link between demand in a customer zone and the quantity ordered from different distribution centres.

Another source of uncertainty can be represented by the transportation times. In the context of long-term decision planning problems this level of detail and the corresponding uncertainty is not relevant and is omitted.

We assume that products are shipped to distribution centres continuously, otherwise excessive quantities are stockpiled at distribution centres. Demands for such products, however, would depend on the market characteristics, with some delivered continuously in the long run while others once in a while. We assume generally that products are sent to customers at a fixed time interval. We then refer to stock that satisfies normal or expected customer demands as operational stock. However, additional stocks are needed to hedge against uncertain demand, to make up for occasions when demand exceeds capacity, to accommodate seasonal demands, to cover demand while other products are being produced and to reduce the impact of major facility failures. Besides, peak stock levels are required for promotional sales. We refer to the sum of these stocks used to hedge against uncertainty as safety stock.

We next focus our attention on production and packing stages and assume that both production and packing lines are located in the same area whereby these two stages can be considered as a single one.

At this stage the demand for a given product p is represented by $\sum_k Q_{tkp}$ so that the quantity q_{tilp} of product p packed by line l at site i must satisfy:

$$\sum_{i,l} q_{tilp} = \sum_{k} Q_{tkp} \qquad \forall\ t,p \tag{3}$$

Thus, denoting by f_{tilp} the coefficient, expressed in *lines/product units*, indicating the usage of a line of type l needed to produce one unit of product p then the planned plant capacity must be at least equal to , expressed in *lines/product units × product units = lines*.

In order to determine the capacity C_{til}, of each plant, note that a line of type l can only operate at a site if and only if the site is open, i.e.

$$z_{ti} \le \sum_{l} r_{til} \le M_i\, z_{ti} \qquad \forall\ t,i \tag{4}$$

If line type l is not in operation then C_{til} is also forced to be 0:

$$r_{til} \le C_{til} \le M_i\, r_{til} \qquad \forall\ t,i,l \tag{5}$$

Then the site capacity C_{til} is given by:

$$C_{til} = C_{t-1,il} + x_{til} - \theta_{til} \qquad \forall\ t,i,l \tag{6}$$

However, in order to obtain C_{til} we must keep track of t', i.e. the time when the expiring investment started. Note that t' depends both on the time τ_{il} required for the capacity of the new investment to be ready for production and on the total operation time T_{il} of such an investment. In real systems such values depend on the technological behaviour of the investment such as the set up time and the life time of the technology. In this case it is necessary to adjust the capacity C_{til} by introducing a capacity adjustment variable C'_{til} such that:

$$C_{til} = C_{t-1,jl} - C'_{til} + x_{til} - \theta_{til} \qquad \forall\ t,i,l \tag{7}$$

where C'_{til} indicates the number of lines reaching the end of their life time at the beginning of the current period t plus the number of lines on which the investment has already been made but that are not yet operating. C'_{til} is a particular feature of our model, it allows us to consider investment with a limited life time investment already done but not yet in operation.

Now we formulate the constraint which links production capacity and ordering quantity as:

$$\sum_{p} f_{tilp}\, q_{tilp} \le e_{til} \left(C_{til} - \sum_{k=t-\tau_{il}}^{t} x_{kil} \right) \qquad \forall\ t,i,l \tag{8}$$

where e_{til} is the efficiency rate of line of type l at site i.

Because the technology used is not perfectly reliable the efficiency rate e_{til} is another source of uncertainty. In fact, in managing production and packing plants the uncertain behaviour of the technological equipment may be taken into account. Machines can be subject to random failures and random repair times. Often efficiency rates e_{til} depend also on the cost of the technological equipment and on the costs allocated for maintenance, but we will not discuss the specific case here.

Finally, another source of uncertainty that must be taken into account when considering long term planning is that due to the economical and political situation introduced in the model by different types of costs.

When a production or packing site is opened for the first time, it is reasonable to assume that there is one-off site opening cost as a result of installing either production or packing lines at the site. Correspondingly, there are site closing down cost savings.

Unit capital costs of lines are assumed fixed regardless of initial or incremental investment. This assumption could be generalized to situations where unit investment costs are different for initial and incremental investments or where unit investment costs vary depending on the number of lines due to price discounts.

The minimum number of units of capacity requirement for initial investment is not represented explicitly to avoid creating a model of large dimensions. On the other hand, the optimal solution would provide some degree of balance between minimum capacity requirement and capital investment cost.

Given the site opening cost v_{ti} of a new site and the saving s_{ti} made by closing a site, the total investment cost on production lines in time t is:

$$\sum_{t,i} \left[v_{ti} \max\left(0, z_{ti} - z_{t-1,i}\right) + s_{ti} \max\left(0, z_{t-1,i} - z_{ti}\right) \right] + \sum_{t,i,l} v'_{til}\, x_{til} \qquad (9)$$

where v'_{til} is the unit capital cost of production line l at site i in time period t.

Costs included in (9) are not the only ones; in order to find the optimal strategy the operational costs are considered as well. Some of the operational costs are fixed and volume independent such as those due to administration and maintenance; some others, such as depreciation, depend on the investment strategy.

Depreciation costs for existing production lines are constants while those related to new investments are calculated as a percentage of the capital cost. Depreciation costs for new investments start to be charged when the investments are ready for production and the charge is made over the depreciation period.

Both the depreciation rate and the depreciation period can be modelled as stochastic variables which affect both constraints and the objective function. Other costs are considered in the model some of them are strongly related to the capacity, like repair and maintenance, or have a stepwise nature, like fixed labour, or are not only capacity related but also product type and product volume related, like set up costs, raw material costs.

All these factors make the problem computationally intractable because of its enormous, and sometimes unrealistic dimensions without even considering the number of scenarios that are used to evaluate a given strategy. This complexity is further compounded by the number of different strategies that must be evaluated, where a strategy is represented by the configuration assigned both to plants and distribution centres.

Because of their importance, strategic planning problems have always received considerable attention and stochastic programming has been a promising methodology for solving such problems. Recently, this field has received growing interest (Bienstock and Shapiro, 1988; Escudero *et al.*, 1993; Holmes, 1994; Malcolm and Zenios, 1994) due to the development of powerful computing technology and optimization solvers (Levkovitz and Mitra, 1993; Maros and Mitra, 1995). A number of stochastic models have been developed in order to make such strategic decisions and in turn evaluate the behaviour of all the feasible strategies, obtained by enumerating the expansion possibilities.

We have developed an optimization approach that identifies a set of 'sub-optimal' strategies which act as starting points for a subsequent refinement procedure. The main advantage of this approach is that it yields a problem of manageable proportions and still considers all the features of the problem by capturing their complex interactions.

3.2   A model instance

To appreciate the complexity of this problem we present the model statistics as shown in Table 10.1.

TABLE 10.1. Model statistics.

Constraints	Real var.	Integer var.	Total var.	Nonzeros	Density
5209	60029	3066	63095	215863	0.07%

This problem is solved on a SUN SPARC20, 128 MBRam, and takes 6000 seconds to solve to a first integer solution. We have made no attempt to solve this problem to optimality.

4   MODELLING UNCERTAINTY

4.1   Representing uncertainty

Our approach to introducing uncertainty in this planning model is based on the creation of a set of scenarios. The uncertainty is represented through a multi-level scenario tree which defines the possible sequence of realizations of the uncertain parameters, called a scenario, over the whole planning horizon, see Figure 10.4. For the purpose of this investigation uncertainty is only incorporated in the demand for products $D_{t,h,p,\omega}$.

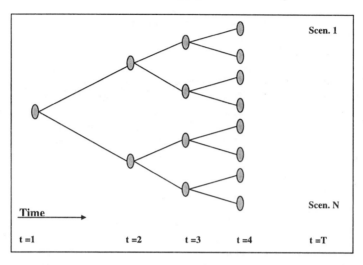

FIGURE 10.4.

With any scenario ω, a probability $p(\omega)$ is associated, such that $\sum_{\omega} p(\omega) = 1$.
Also the probabilities are independent from period to period, and the quantity demanded with scenario ω in year t does not need to bear any particular relationship to the demand with scenario in the year $t+1$. Thus we obtain a stochastic programming model .

In stochastic programming, a lot of emphasis is placed on the decision to be made today, given present resources, future uncertainties and possible recourse actions in the future. The uncertainty is represented through a scenario tree and an objective function is chosen associated with the sequence of decisions to be made and the problem is then solved as a large scale linear program. Of course, in most situations the number of possible scenarios is so great that it is impossible to model them explicitly since this leads to a linear program of enormous size. For this reason stochastic models address shortcomings of the deterministic models but have not been widely used so far because they are intractable in size. The success of stochastic programming depends on the extent to which a relatively simple scenario tree can be constructed that captures the risk inherent in making one decision today. In many such cases, practitioners often employ a technique called 'scenario analysis' whereby the uncertainty is modeled via a few scenarios.

4.2 Scenario analysis

In scenario analysis given a finite set of scenarios $s = 1, ..., S$ which represent different realizations of all the uncertain parameters, the model is solved for each scenario. The aggregated solutions obtained represent an important information source for the management. They are heuristically analysed as a whole by the planners who can adapt the method to the case and investigate the most useful aspects of the problem. In this way we have a complete overview of all the possibilities that can occur and can evaluate the related consequences, this gives us a deeper insight into the problem. We then use this information to directly attempt to find a better solution or incorporate it into other problem solving approaches.

By comparing these solutions we are able to verify the stochasticity of the problem and similarities among the solutions can be found. The validity of the single solutions can be tested against all the scenarios checking their ability to cope with every scenario occurrence.

4.3   Hamming distance and solution matching

After solving the model for each scenario, we analyse both scenarios and the related configurations. A configuration s is the strategic solution obtained solving the model for the particular scenario ω.

An interesting aspect to look at is whether some of the scenarios lead to the same strategic decision; in this case the configurations are identical and it would be possible to consider just one of them eliminating the other, thus reducing the number of possible decisions and consequently the complexity of the problem.

This analysis can be done by evaluating the *hamming distance* between each pair of configurations. The hamming distance is defined as the number of differences that exist between two strategic decisions. We create an hamming matrix $A[i,j]$ (triangular matrix) where $a[i,j]$ is the hamming distance between configuration i and configuration j.

At the same time *solution matching* is carried out over all integer solutions which verifies whether some of the solution values are identical in all the configurations. if this happens it means that the particular decision may be fixed for every scenario occurrence. The variable can then be considered as data in the model and fixed to the value found, reducing the number of variables in the model thus simplifying the solution process.

Analysing the strategic solutions is useful to measure the problem stochasticity. If the configurations are similar for every scenario it means that the scenario variance does not affect the model solution which can then be considered robust enough for every scenario occurrence. Moreover, always observing the integer solutions, it is possible to find out information about the scenarios such as which are the extreme scenarios and which others are similar. This information can be used as a good starting basis for further analysis methods.

4.4   Configurations analysis against scenarios

First we analyse how each of the strategic configurations hedges against all the scenario occurrences.

This approach is a *two stage* method where in the first stage the model is solved for all scenarios {1,...,W} thus obtaining a corresponding number of configurations; this number can be reduced to K when some of the configurations are found identical by the hamming function. In the second stage each configuration is evaluated against all scenarios. Suppose that decision k is made, the corresponding configuration k is fixed in the model which is then repeatedly solved for all the scenario occurrences. In our model the strategic variables are integer and do not depend on the scenario, while the operative variables are of real type and depend on the demand scenario. Once the configuration is fixed (integer variables fixed) the model to solve in the second stage of the method is a Linear Programming (LP) model; the method is illustrated in Figure 10.5.The optimum value of the LP model solved for scenario j and configuration i fixed, represents the effect that configuration i has if scenario j occurs. For instance if the decision taken is ideal for an extreme scenario with a low demand level, the capacity level set up is correspondingly low, if a high demand scenario occurs the company is not able to meet the demand thus suffering high shortage costs, this situation is probable to happen for every extreme configuration.

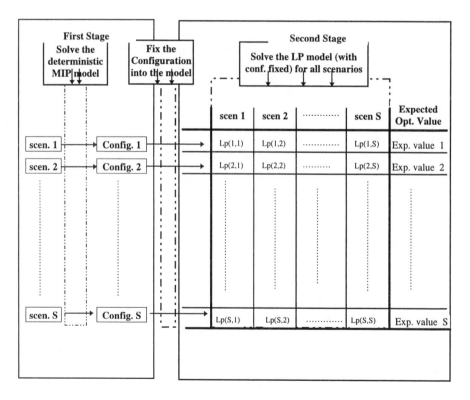

FIGURE 10.5. Configurations analysis against scenarios.

The average of the LP optimal values of each configuration gives an indication of the decision behaviour over all scenarios. It is also interesting evaluating the best and worst values whose difference indicates the risk of the decision, it may give a high profit for one or more scenarios but if some other scenarios occur the effect could be devastating for the company. The best expected value corresponds to the configuration which best reacts to market variations.

Our next investigation of the results obtained is to establish the best configuration for each scenario, that is the configuration corresponding to the best value of each scenario column in the table represented in Figure 10.5. The validity of such an approach is because the model cannot be solved to integer optimality but only to the first integer solution. This implies that only some of the configurations found are relevant since this analysis can highlight a better configuration than those initially found as the first integer solution of the model for each scenario. This analysis would not apply if the model could be solved to optimality.

5   RESULTS

The scenario analysis techniques introduced in section 4 have been applied to a capacity planning model.

The only uncertain component in the problem is represented by the demand for products. 100 different demand scenarios have been supplied by the company.

In order to solve the deterministic model for each scenario we have created a procedure to carry out this in batch mode. This procedure starts by reading in an MPS file, and then from run to run it updates the right hand side vector with the appropriate demand. Figure 10.6 shows the objective values of the solution obtained, presented in a chart.

Our first analysis on the results obtained from the previous procedure consists of calculating the hamming distance; an appropriate procedure has been developed which compares each pair of result files, and creates a triangular matrix whose element $a_{i,j}$ is zero if the solutions i and j are identical.

```
Read MPS file ('parameters')
For c from 1 to 100
        read in configuration
        modify 'parameters' (bounds)
        For ω from 1 to 100
                read demand
                modify "parameter" (demand)
                call Optimizer
                return the solution optimal value.
        endfor
endfor
```

FIGURE 10.6.

The same procedure carries out the matching function as well. A special value is set for those variables that have different values in different solutions.

These two procedures are mainly used to verify the difference between scenarios thus identifying the model's stochasticity.

Finally, scenario analysis is carried out on the results to assess how each configuration fairs against each scenario. We have created a function for fixing the appropriate bounds of the integer variables (i.e. the values for the current configuration under investigation) prior to optimization. Exploiting the features of the solver and of the problem the function is carried out through the steps shown in Figure 10.6.

The output from this procedure is a 100×100 matrix of optimal LP values. Various analyses is then carried out on these results. An average as well as the best and worst optimal value is identified for each configuration, which is displayed in Figure 10.7. The best average value corresponds to the configuration that behaves the best for all the scenario occurrences. In Figure 10.8 a graphic representation of the maximum revenue obtainable under the various scenarios is shown.

Comparing Figures 10.7 and 10.8, it is interesting to notice that the best configurations are usually generated under large demand scenarios, we can then assume that, in this application, it is preferable to set up a large production capacity and pay high maintenance costs rather than suffer shortage costs which are much higher.

Furthermore, it is also possible to identify which configuration is the best for each scenario. This analysis is necessary since we do not solve the model to integer optimality but instead stop at good integer solution. In this analysis, we find that some of the configurations are better for more than one scenario; these results are given in Figure 10.9 from which we can see that only 15 configurations are effectively relevant for the problem and these need only be considered as possible candidates for the strategic decision.

6 CONCLUSION

A manufacturing supply chain problem has been investigated in this chapter. It consists of a capacity planning problem over a long time horizon. It has been established that making strategic decisions has to incorporate the operational decisions that are made in the short term. In order to consider all the addressed costs the usage of corporate and analytic databases have to be accessed at the modelling stage. By combining the strategic decisions and operational decisions which traditionally are treated as two separate models, results in the development of large MIP models. This approach is known as integrated logistic optimization modelling.

A complete formulation of the model under investigation and different aspects of the problem solution have been analysed. We model uncertainty through a scenario approach and analyse different stochastic solution methods.

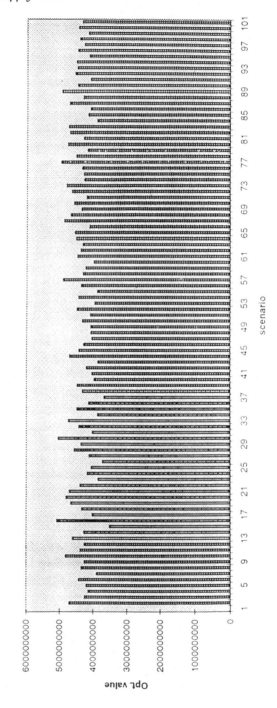

FIGURE 10.7. Integer optimal value for each scenario.

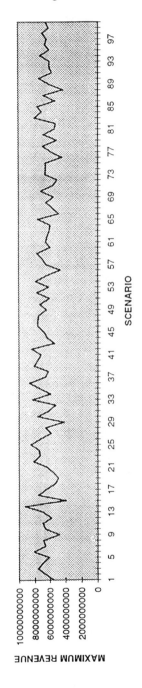

FIGURE 10.8. Scenario's maximum revenue.

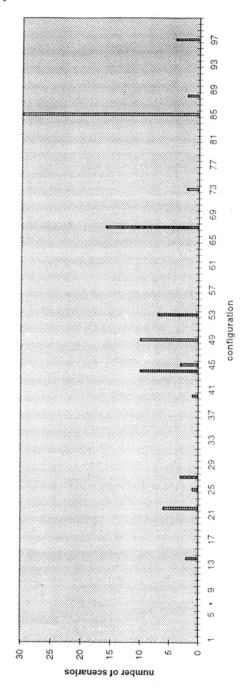

FIGURE 10.9. Number of scenarios for which the conf. is best.

In this context the concept of optimal solution loses its validity, since optimality depends effectively on future parameter realizations which are uncertain. In an environment under uncertainty it is important to achieve a good solution able to hedge against all the scenario occurrences. The existing methods can be mainly divided into proactive and reactive approaches. In a reactive approach useful information are provided by the analysis of optimal solutions to different realizations of the uncertain parameters. This information provides an insight to the decision maker to the consequences of making certain decisions. For instance, applying the scenario analysis to the problem under investigation identified 15 key configurations out of 100. It also calculated the risk of each decision and their overall performance against all scenarios. From the results we observed that the best configurations were obtained from scenarios with an high level of demand, this leads to the conclusion that it is preferable to install a large capacity configuration since the shortage cost entails more risk than the cost due to overcapacity. This method even if not precise provides management with good indicators on making strategic decisions.

Proactive methods involve developing models which iteratively yield good solutions taking into account uncertainty. This involves a much larger model which usually assumes huge dimensions, for this reason it becomes difficult to apply these methods to real life problems which are in general very large. To this end, we have found that the scenario analysis acts as a good starting point for finding solutions that take into account uncertainty.

REFERENCES

Bienstock, D. and Shapiro, J. F. (1988) Optimizing Resource Acquisition Decisions By Stochastic Programming, *Management Science*, Vol 34, No 2, pp. 216-229.

Birge, J. R. Decomposition and Partitioning Methods for Multi-Stage Stochastic linear Programming, *Operation Research*, Vol 33, pp. 989-1007.

Codd, E. F. and Codd, S. B. (1994) OLAP with TM/1, E.F. Codd and Associate.

Codd, E. F., Codd, S. B. and Sally, C. T. (1993) Providing On-line Analytical Processing to User Analysts: An IT Mandate, E.F. Codd and Associates.

Cohen, M., Kamesam, P. V., Kleindorfer, P., Lee, H. and Tekerian, A. (1990) Optimizer: IBM's Multi-Echelon Inventory System for Managing Service Logistics, *Interfaces*, Vol 20, pp. 65-82.

Davis, T. (1993) Effective Supply Chain Management, *Sloan Management Review*, Summer, pp. 35-46.

Dempster, M. A. H. and Thompson, R. (1996) EVPI-Based Importance Sampling Solution Architectures, to appear in: *Annals of Operations Research*, Proceeding of the POC96 Conference, Versailles, March.

Eppen, G. D., Martin, R. K. and Schrage, L. (1989) A Scenario Approach to Capacity Planning, *Operations Research*, Vol 37, No 4, pp. 517-527.

Escudero, L., Kamesam, P. V., King, A. and Wets, J-B. (1993) Production Planning via Scenario Modeling, *Annals of Operations Research*, Vol 43, pp. 311-335.

Holmes, D. (1994) A collection of stochastic programming problems, Technical Report 94-11, Department of Industrial and Operation Engineering, University of Michigan, Ann Arbor, MI 48109-2110.

King, A. (1994) SP/OSL Version 1.0 Stochastic Programming Interface Library: User's Guide, Research Report RC 19757 (87525) 9/26/94 Mathematics, T.J. Watson Research Centre, Yorktown Height, N.Y.

Levkovitz, R. and Mitra, G. (1993) Solution of Large Scale Linear Programs: a Review of Hardware, Software and Algorithmic Issues, in: T. Ciriani and Leachman (Eds.) *Optimization in Industry*, John Wiley, pp. 139-172.

Lucas, C., Messina, E. and Mitra, G. (1996) Risk and Return Analysis of a Multi-Period Strategic Planning Problem, in: A.H. Christer, S. Osaki and L.C. Thomas (Eds.) *Stochastic Modelling in Innovative Manufacturing*, pp. 81-96.

Malcolm, S. A. and Zenios, S. A. (1994) Robust Optimization for Power Systems Capacity Expansion Under Uncertainty, *Journal of Operational Research Society*, Vol 45, No 9, pp. 1040-1049.

Markowitz, H. (1959) *Portfolio Selection*, Yale University Press, New Haven, Conn.

Maros, I. and Mitra, G. (1996) Simplex Algorithms, in: J. Beasley (Ed.) *Recent Advances in Linear and Integer Programming*, Oxford University Press, pp. 1-46.

Mulvey, J. M., Vanderbei, R. J. and Zenios, S. A. (1995) Robust Optimization of large-scale Systems, *Operations Research*, Vol 43, No 2.

Sach, T., Collins, I. and Page, J. (1996) The Data Warehouse Report, UNICOM Information Technology Reports.

Schoemaker, P. J. H. (1988) The scenario approach to strategic thinking: methodological, cognitive and organizational perspectives, Centre for Decision Research, Graduate school of Business, University of Chicago (February).

Shapiro, J. F. (1993) The Decision Database, Sloan School of Management Working Paper #3570-93-MSA.

Shapiro, J. F., Singhal, V. M. and Wagner, S. N. (1992) Optimizing the Value Chain, *Interfaces*, Vol 22, No 5.

Sims, M. J. (1991) Use of a stochastic capacity planning model to find the optimal level of flexibility for a manufacturing system, Senior Design Project, Department of Industrial and Operations Engineering, University of Michigan, Ann Arbor, MI 48109.

Stockdale, R., Hammett, A. and Barrie, S. (1997) Data Warehousing: Tools and Solutions, Bloor Research.

11 New Trends in Rapid Response Manufacturing Logistics

P. Dempsey
Nossmole Dempsey and Company Ltd.

1 INTRODUCTION

Benetton, the Italian sportswear manufacturer, distributes 50 million pieces of clothing worldwide through a highly efficient Quick Response linkage among manufacturing, warehousing, sales and the retailers they serve. When, for example, a popular style of winter-weight slacks is about to sell out early in the fall, a retailer records the product through direct links with Benetton's mainframe computer in Italy. The order is downloaded to an automated machine that makes the slacks in the necessary range of sizes and colours. Workers pack the order in a bar-coded carton addressed to the retail store, and then send the box to Benetton's single warehouse – a highly automated $30 million facility shipping 230,000 pieces of clothing a day to 5000 stores in 60 countries. Including manufacturing time, Benetton can ship the completed order in only four weeks.

In 1988 **Nissan** recognized that 30% of customers ordering vehicles from the auto industry settled for a car that was available rather than the precise specification they wanted. Nissan responded with the SOMO project (sell one make one), the object of which was to supply within two weeks of order any one of around 35000 variants. Now other competitors are looking at ways of producing the '4 day car'

These are just two examples of the market trend that is forcing rapid response manufacturing into the spotlight as a key competitive requirement of world class performance companies. The trend is towards greater customer focus. This means not only short lead time for delivery but minimum inventory which in itself demands rapid response not only in the manufacturing area but in every element of the supply – manufacture – distribution chain.

The result of Nissan's decision was effectively to change the way they do business. The need for rapid response pervaded every aspect of and department in their business.

It is important to recognize, therefore, that there is no such thing as a Rapid Response Manufacturing Logistics Project (RRML) or initiative; RRML is part of a continuum of activity from concept of a product through its design and manufacturing cycle to distribution and post sales service.

This chapter is therefore organized to describe the trends in some aspects of that continuum, namely:

- The Supply Chain
- Concurrent Engineering
- Organization and Layout
- Automated Storage and Retrieval
- Cross Docking
- Warehouse Management
- Distribution – Vehicle Management
- Customer Order Processing
- Demand Planning.

It will also describe trends in three of the most important enabling technologies or levers in the continuum:

- Electronic Data Interchange
- Client Server Technology
- Expert Systems.

TABLE 11.1.

Plant Performance % Change 1992 to 1994	UK Suppliers to Japanese transplants	Other UK Plants	Overall	Japanese Plants
Units per hour	+ 37.8%	+ 19.1%	+ 31.5%	+ 38.4%
Production volumes	+ 54.7%	- 4.1%	+ 35.1%	- 15.8%
Value of sales	+ 41.1%	+ 4.7%	+ 35.7%	- 18.0%
Factory headcount	+ 27.4%	+ 4.0%	+ 19.6%	- 26.2%
Stock turn ratio	+ 187.4%	+157.5%	+ 177.4%	- 15.3%

Finally it will summarize the constraints upon the adoption of these trends and the likely rate of adoption of them.

What is certain is that global competitive pressure for improved performance in manufacturing is relentless and time based. For example, a 1994 worldwide manufacturing competitiveness study by Andersen Consulting concluded that the UK/Japan gap is widening. Comparing Japan and the UK in 1992 and 1994, Japan is clearly improving significantly, despite the recession. Major restructuring is occurring. Although requests for cost-reduction were modest in 1993, cost-reduction targets are 20% to 30% for the period 1994 to 1996. Product simplification at the design stage and greater parts standardization are under consideration.

The UK also shows big improvements in performance (Table 11.1), but these are not as great as in Japan. It is also unclear how much of the UK improvement is due to volume increases rather than genuine process improvements, and therefore how long this can be sustained.

2 TRENDS IN SUPPLY CHAIN MANAGEMENT

Measurement of any operation is clearly an essential step in controlling its quality and efficiency. However, few companies yet measure the overall efficiency of the supply chain, allowing trade-offs to be made between different supply and distribution options. Direct product companies use them for evaluating the supply chain to balance costs against service level, and as such they are progressive tools. However, Direct Product Profitability (DPP) models' primary purpose is to compare products rather than distribution efficiency or options. They often omit some distribution costs while including other costs of a different nature (e.g. store labour).

Pioneering systems within the supply chain management area have a number of common principles. For example, each level within the supply chain uses consistent planning tools and processes. Specifically, tools used should include time phased replenishment tools such as Distribution Requirement Planning (DRP) and Manufacturing Requirements Planning (MRP).

The rapid evolution of the supply chain has led to the requirement to have an end to end view of the supply chain. The degree of control demanded is similarly being set within tight constraints, since it is not sufficient to simply fulfil orders within a given service level or fill rate. Tracking and tracing down to pack or batch levels is required so that stock visibility is provided across the supply process.

The information systems used must therefore, be integrated and capable of pushing demand from one level to the next (e.g. from sales to distribution, or manufacturer to supplier). A demand for an end product at one level may become a demand for various raw materials at the next level.

Integrated systems also allow quick transmission of information about changes in demand and the plans to meet demand. Additional features of such systems include the ability to promise customer delivery dates based on production or configuration scheduling, tracking customer shipments through the entire pipeline and ensuring sales forecasts can be translated into realistic production schedules. This integration can be extended to other companies in the supply chain, such as customers and suppliers, to achieve greater improvements in logistic performance. A key component of this integration is the effective implementation of Electronic Data Interchange (EDI) and this is discussed in greater detail in the 'EDI section'.

Pioneering systems have a greater focus on planning and decision support tools that enable the company to make the best operational decisions across the enterprise. However, many companies are focusing on the integration of their supply chains at the transaction system level. Such as inventory control, order processing and shop floor management. Although these systems are critical to a company's day-to-day business because they manage the minutiae of operations, they provide very little planning and communication between the functional groups of the business, such as marketing, distribution, manufacturing and purchasing. They do not assist in making operational decisions but focus on the current status of the business (See Figure 11.1).

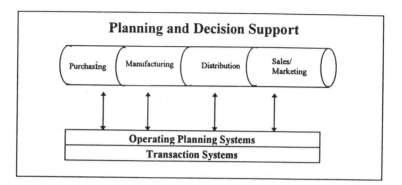

FIGURE 11.1. Planning and decision support.

Planning and decision support systems are focused on achieving logistics integration through better cross-functional communication and better visibility into all the business's operations, thereby enabling informed operational decisions and lower costs. They achieve this integration by extracting the data available in transaction systems. They take the key data from each of the primary transaction systems needed to give a complete, integrated view of the company's distribution, manufacturing and purchasing functions. Their power lies in the aggregation of the key data needed to run the business, collected in one place and available to all decision makers. These systems also provide other tools such as graphics and simulation facilities to enhance usability and performance.

Pioneering systems focus on evaluating the total delivered cost of a product throughout the entire supply chain, from purchasing through to distribution, against a target service level, allowing performance to be measured and opportunities for operational improvements to be identified. The total supply chain performance is measured according to a set of seven or eight indicators, so that a 'balanced scorecard' is achieved. These indicators are often based on the following criteria:

- **Financial:** product profitability, total delivered cost, reliability of performance.
- **Customer focused:** delivery time, accuracy against customer requirements, non-adversarial relationship, supplier performance tracking.
- **Internal:** innovation, ability to shape customer requirement, quality service, core competencies.
- **Growth:** continuous improvement, product and service innovation, empowered work force.

It is essential for these indicators to be made quantifiable so that performance can be measured over a period of time. Since vast amounts of data will be needed

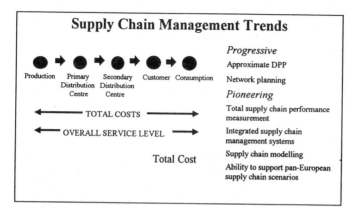

FIGURE 11.2. Supply chain management trends.

for supply chain performance measurement, it should be possible for systems to gather all the required information at a high level and make provisions for drill-down to more detailed levels (see Figure 11.2).

Supply chain modelling packages use the above indicators to develop and test different supply chain scenarios. Using actual historic orders and repeating the simulation with different rules and parameters allows many scenarios to be tested without affecting the live environment. Specific applicators for supply chain modelling packages can be in the areas of stockholding and transport between multiple European warehouses, choice of forecasting algorithms and assessment partnerships.

Progressive features within supply chain modelling packages are regarded as:

- Prompting the user to enter parameters such as locations of warehouses, customer locations, characteristics of transport links etc. The system then determines the optimum location of new warehouses or the optimum transport link to be used. Simulation features are provided to determine the optimum configuration of sales territories, effects of pooling resources with competitor to rationalize facilities and the cost of providing new services such as providing tighter delivery windows.
- The implementation of long term resource planning to assist depot management in assessing resource requirements (vehicles, drivers, warehouse staff etc.) against required service levels, and development of a business case for adjusting any operational variables, e.g. decommissioning vehicles, altering the driver profile or changing branch schedules.

Many firms are either contemplating or implementing a pan-European strategy, which requires greater co-ordination across the entire logistics pipeline and across all European countries. Thus, the supply chain requires reconfiguring to manage supply and demand on a pan-European basis. This can involve focused factories that fulfil orders for one type of product for several countries rather than on a purely national level. The ability to support these pan-European supply chain scenarios is regarded as pioneering.

3   TRENDS IN CONCURRENT ENGINEERING

Concurrent engineering is a key issue in world class business practice because it reduces time to market of a new product while reducing life cycle costs and improving quality. In particular it reduces life cycle costs to the programme because normally 75% of these costs are committed in the early stages of design.

George Hess, Vice President of Ingersoll Milling who have won national recognition for their applications of advanced technology to manufacturing says *'Goals, targets and performance must be shared with suppliers and stakeholders very early in the program and must be centred on customer satisfaction – not just product performance, cost or even quality. Those are supporting objectives that help us meet our primary objective of customer satisfaction, but they are not in themselves customer satisfaction. Customer satisfaction and that alone is what customers buy and tell their friends about. That is the breeding ground of real success.'*

As a trend setter in the area of concurrent engineering Hess advocates the chronology and key activities of a concurrent engineering project as follows.

CHRONOLOGY OF A CONCURRENT ENGINEERING PROJECT
Concurrent engineering projects usually involve a production program to introduce a new product. The phases of this type of project are:

- Phase # 1 – Technology and concept development
- Phase # 2 – Product/ process development and prototype
- Phase # 3 – Process validation and product design confirmation
- Phase # 4 – Production and continuous improvement.

The key activities in a new product introduction program are:

- Prepare clearly stated overall project objectives
- Review product plan and design objectives
- Develop real team spirit between product and process designers
- Complete the initial product design and documentation
- Identify design changes to improve manufacturability
- Integrate material suppliers as effective team members
- Plan practical level of flexibility
- Finalise product design
- Finalise optimum process design for planned production level
- Develop factory layout and manning plan
- Simulate manufacturing process
- Prepare costs plan

- Continue full team involvement during design and build phase
- Benchmark against the best
- Continuously improve the product and process design.

Aerospace automotive and defence industries in general set the trends for concurrent engineering. Typical of these are:

- Paperless design to manufacture using rules based feature analysis of a solid model object oriented technology and driving STEP implementation wherever possible.
- Seamless technology that permits the lifetime costs of a component or product to be determined long before the component reaches the factory floor.
- Rapid prototyping of various types and graphical techniques approaching virtual reality in the design for assembly area.

The opportunity for improvement in competitive performance through concurrent engineering are almost endless and one of the most important development areas in manufacturing business.

4   TRENDS IN ORGANIZATION AND LAYOUT

Factory organization and layout are often treated as the poor relations to strategy, finance and technology. Yet the way a factory is organized and laid out is at the root of workflow and therefore cashflow and ultimately profitability. Of equal importance in these days of awareness of the impact of the work place on motivation, the layout of the factory can have a major influence on the behaviour of the people who work in it. Finally since rapid response is an essential element of customer service, the need for flexibility to change layout is greater than ever before (Figure 11.3).

In the light of these issues current key trends in factory organization and layout include:

ORGANIZATION AND LAYOUT HAS A DIRECT EFFECT
ON CUSTOMER AND SHAREHOLDER VALUE

- Cost (space, transport, services, labour, inventory)
- Time (delivery, response, flexibility)
- Quality (right product, right place, right time)
- Image
- Supply Chain (delivery, distribution, compatibility)
- Staff motivation

FIGURE 11.3.

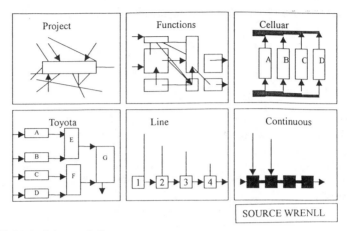

FIGURE 11.4. Material flow patterns.

- Much greater emphasis and care about the choice of manufacturing location in both global and national aspects. The location of factories relative to the markets they serve is a trade off between a vast array of political, financial, cultural, logistical, technical and commercial considerations. The complexity is immense as is the opportunity to get it wrong. These choices can govern the ultimate success of a business. Sophisticated software is available to help solve these problems and the trend is towards its greater use.
- The local infrastructure at a plant location has also become a prime focus for attention. The trend is for concern not only, for example, whatever the local planning authorities can and will co-operate with buildings and local road ways but also the impact the plant has on the local environment. A major oil company recently offered to purchase every house in a village near its plant in Wales. This was not because it wished to enter the property business but because it believed its refinery was detrimental to the environment. Its European Chief Executive was known to have expressed the view that by the turn of the century a company's actions and behaviour will be as important as its products.
- The use of systematic analysis to identify the best groups of products to make together. For example, Pareto analysis to identify the volume/variety/value split of production so that each element can be laid out to suit best its process requirements.
- The use of a firm methodology for laying out manufacture of each group of products (see Figures 11.4 and 11.5).
- Recognition that the labour force spends a large proportion of its life in the factory and should be involved in the planning process in at least two ways; assessment of the degree to which a layout provides motivational incentives like autonomy, responsibility, freedom of movement, visual stimulation and communication; participation in testing and suggesting layout ideas.

METHODOLOGY IS VITAL

- Define the operation sequence and material flow
- Do proximity analysis
- Do movement frequency analysis
- Establish individual space requirements
- Do space relationship diagram
- Establish initial block layout alternatives
- Modify and choose based on common sense and practical issues
- Apply detail and finalise

FIGURE 11.5. Methodology is vital.
Source: Wrennall, B. *Handbook of Commercial & Industrial Facilities Planning.*

- Flexibility through avoidance of immovable features in the floor space available. Equipment with deep foundations, electrical substations, service plant and all such items are organized away from the main work area either outside, over head or under ground. Progressive heavy machine makers like Ingersoll Milling for example, have developed machines which no longer need foundations, costing millions of dollars and which are totally immovable.

Organization and layout is a serious activity which directly affects customer, shareholder and stakeholder value. Often companies guess at the options and make costly mistakes which last the lifetime of the plant. William Wrennall points out quite correctly that facility layout is the key to workflow, cashflow and profit.

5 TRENDS IN AUTOMATED STORAGE AND RETRIEVAL

The three main elements of an automated storage and retrieval system are the stacker cranes that move the pallets and store them in the racks, the racking itself and the control system that manages the movement of stock. The cost of both control systems and stacker cranes has decreased by at least 50% in recent years, while functionality in both areas has increased. This narrowing cost differential between automated systems and traditional fork lift trucks has resulted in the emergence of a new generation of low cost automated systems (see Figure 11.6).

The level of automation can encompass the storage, retrieval and movement of raw materials, packing, work in progress and finished goods. An example of this is the movement of pallets via complex computer controlled conveyor systems from the factory or receiving dock to the warehouse. The system should have some level of intelligence built in so that, for example, fast moving products are stored at the nearest available location for easy access. Exception checking should

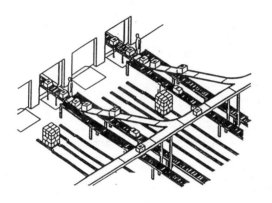

Progressive

Automated warehouses for
storage, retrieval and
movement of goods

Pioneering

Interactive computer link to
overall systems

Automated order picking for
individual units

Automated cross-docking

FIGURE 11.6.   Automated storage and retrieval systems.

also be a feature, so that if a pallet is too high, too wide or too long, it is automatically set aside for reloading. Another example of automation is the use of robotic technology to store and retrieve items from high bay storage areas, so that picking efficiency is further improved. One or two aisle automated pallet stores using modern stacker cranes, in comparison to fork-lift trucks, do not require super flat floors, have low operating costs, are mechanically simpler and thus more reliable. Finally, systems to support automated warehouses should be able to handle a multishift operation in order to reap the benefits of reduced manning.

Although there are increasing examples of minute to minute interactive links between the PLCs (programmable logic controllers) managing the materials handling equipment and the warehouse management system running on the main computer, these situations are still unusual and are considered to be pioneering.

A potential for increased warehouse automation lies in automated order picking systems, especially those for the picking of individual units from a wide variation of products rather than layers of pallets. Another new area generating substantial interest in the fast moving goods arena is the automation of cross docking, and this is covered in the next section.

6   CROSS DOCKING

Cross docking can be defined as any method that changes the sequence or quantity of supply of a product through a factory or distribution system without putting the product into storage.

It has evolved much interest in the food distribution industry in recent years especially food retailing. It is also, for example, a key feature of the ground breaking PWAF (Plant With A Future) manufacturing plant renewal programmes

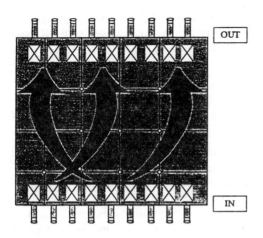

Pioneering

Support of three main
types of cross-docking

Dynamic scheduling

Rapid data capture

Flexible stock
accounting

FIGURE 11.7.

undertaken by Caterpillar in its plants around the world. The systems required to support these initiatives are certainly pioneering and cut across many functional areas. The features incorporated in systems to support cross docking will include the following (see Figure 11.7):

- Support of the three main types of cross docking:
  - Full pallet moves directly from inbound truck to outbound truck with the product requiring no extra preparation or sortation.
  - Case or pallet sortation moves a product through the distribution centre sorting for a specific store or machine location.
  - Goods requiring additional preparation are moved through the distribution centre and additional activities are performed.
- Dynamic vehicle and load scheduling which co-ordinates the arrival and departure of incoming and outgoing vehicles. Optimum use of each mode of transport should be made in order to achieve the greatest transport economy. This may be further complicated if the incoming and outgoing vehicles cross company boundaries. In compiling the schedules the system may make use of standard timed for vehicle loading and unloading.
- Stock should now flow through the cross-dock rapidly and should not be hindered by the manual recording of stock movements. Stock should also be tracked through all stages of storage, handling and transport. Consequently the system should support rapid data capture and may feature barcode technology, truck mounted terminals or tagging technology.
- Traditional stock counting and control mechanisms can be difficult to apply to most cross docking scenarios, and systems must have additional flexibility to deal with shortfalls and overstock.

For those particularly interested in this subject the airlines with their check-in and baggage handling facilities are probably the most tried and tested of all cross-docking systems, without actually being recognized as such. They are becoming highly automated now but there are many low technology lessons to be learned for use in manufacturing which have been common practice at airports for years.

7   WAREHOUSE MANAGEMENT

At first sight the concept of a warehouse as part of rapid response manufacturing logistics would appear alien. The cross docking arrangement already described are indeed the most rapid form of sequencing without the need for storage.

Realistically, however, there is a real need for warehousing within the continuum of activities making up rapid response manufacturing logistics. The need derives from many sources, not the least being the uncertainty of transport times, so often affected by weather and traffic. But also problems arise from supplier errors or features in load quantity and forecast errors.

Given that warehouses have to exist, through advances in hardware and software technology they have become a hot bed of advanced technological applications. Quite commonplace these days are hand held computers for data collection, voice activated systems and tagging technologies to track the flow of pallets or goods through the warehouse. Warehouse management covers operational areas such as picking, put away and warehouse layout. At this point in time it is reasonable to expect warehouse management to support the following functions, and we would consider these to be progressive (see Figure 11.8).

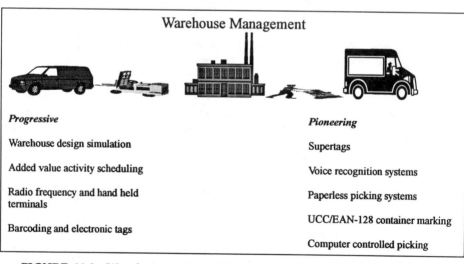

Warehouse Management

Progressive

Warehouse design simulation

Added value activity scheduling

Radio frequency and hand held terminals

Barcoding and electronic tags

Pioneering

Supertags

Voice recognition systems

Paperless picking systems

UCC/EAN-128 container marking

Computer controlled picking

FIGURE 11.8. Warehouse management.

- The use of computer simulation programmes for planning and selecting the optimal warehouse layout. This allows warehouse managers to evaluate various warehouse design alternatives, for example block stacking or drive-in racking, design of picking area, number of counterbalance, reach for forklift trucks etc. The use of this type of tool is increasing as a greater focus is placed on ensuring that warehouses are used to their maximum efficiency and capacity.
- The scheduling and control of 'added-value' activities like cutting, packing, kitting, personalization and customization is also becoming increasingly automated, which is especially valuable where these activities lie on the critical path between picking and dispatching goods.
- The use of R.F. (radio frequency) communication between the warehouse system and terminals mounted on mobile equipment or hand held terminals, eliminating paper, reducing errors and improving productivity. One of the main advantages of using radio frequency systems as compared to other approaches is that they do not require line-of-sight between the tag and the reader. This means that they work in environments with excessive dirt, dust, moisture and poor visibility. Often barcodes and the appropriate laser scanning devices are combined with R.F. or hand held terminals to further improve the productivity and accuracy of item put away, order picking, product movement and stock checking.
- Barcoding introduces the possibility of more detailed tracking of products, from initial receipt to the warehouse through storage and picking to final delivery to a customer. Two dimensional barcoding is increasingly being used and this holds additional data such as warnings, storage details and disposal instructions for medical goods, hazardous waste and perishable products. This improves the speed, accuracy and quantity of data collection, and thus increases the speed of moving goods. A problem which prevents more widespread use of barcodes in Europe is the lack of a widely implemented coherent article numbering convention. Barcodes are often not unique, and so the same article from different manufacturers may have two different barcodes.
- Electronic tags (transponders) can be embedded in or attached to individual objects or product carriers (pallets, boxes) to control, detect and track without the need for a human operator. When a pallet arrives at a distribution centre, the unique ID code of each pallet is read and it is weighed and inspected. This information is captured in a database. Empty pallets sent back to suppliers are also registered automatically as they leave the distribution centre. Automatic identification allows the warehouse personnel to maintain accurate shipment records, track all pallets in circulation, analyse the lifecycle of pallets and reduce time. The data collected can be either sent directly to a host computer or programmable logic controller, or it can be stored in a portable reader and later uploaded to a computer.

Some companies with large distribution operations, primarily in the food, consumer products and electronic industries, have been pioneering in the following areas:

- Super tags are seen as replacement for barcodes in warehousing applications. This is possible due to the development of a low cost transistorized radio transponder device using a single integrated circuit and a printed flat aerial. A scanning device can read a whole collection of these codes by remote control at one pass, thus allowing components in a bin or trolley to pass through stock control checks without the need for each item to be scanned individually.
- Voice recognition technology has now developed to the point where it is being used in warehousing environments. Voice recognition systems are used to collect and enter data from remote locations so that the operator's hands and eyes are free to carry out other tasks. Voice recognition is being used for receiving, order picking, inventory control and quality inspection, and benefits are cited as increased productivity, greater efficiency and improved data accuracy. For example, incoming products can be quickly entered into a company's inventory control system by verbally entering the relevant information in the system's database regarding the shipment is automatically reconciled and updated to reflect the receipt. Any necessary labelling and documents are automatically generated to identify and describe the products.
- Paperless picking systems are focused on the automation of the picking process when assembling customer orders. The typical arrangement is that the storage area (flow rack, shelving etc.) is divided into fixed zones. Each picker has a number of zones and either receives a container or starts a new container to pick into. Once the picking system knows the order or container number it immediately begins to sequence the quantities and lights for picking. When the pick is complete all lights are out, the zone controller or hand-held terminal indicates that it is ready for the next order and the container is sent to the next picker or to dispatch.
- Although several methods of container marketing currently exist, competing technologies to the Universal Container Code will only fill niche roles for the next ten years while the UCC/EAN-128 becomes the dominant code. The UCC/EAN-128 is able to add value in terms of providing information about:

  - Weight or other variable measure that can be used to produce more accurate invoices for the supplier
  - Tracking products more accurately where lot information is critical (e.g. pharmaceuticals)
  - Tracking information about high-value products throughout the supply chain
  - Product freshness where perishability is key.

As an example, it is predicted that this marketing technique will be used for around 50% of all grocery cases, around 80% of general retail cases and all pallet and truckload containers.

Computer-controlled picking vehicles run on a track. Picking requests are communicated via infra-red or R.F. at the start of the pick cycle. Up to ten orders may be picked on one circuit. The picker stands on the vehicle. The vehicle travels to the first location and tells the picker what to pick, where to pick it from and which tote to put it in. The system weighs the pick and, if correct, moves to the next location. Supervisor override is allowed for pick discrepancies. When it returns to the start of the cycle, the system communicates the total pick.

8   DISTRIBUTION – VEHICLE MANAGEMENT

The real value of rapid response manufacture is not achieved until the product is accepted by the customer. The distribution or delivery system to the customer is therefore vital. For example, one forklift truck manufacturer in the late 1980s had done the work in his factory to reduce manufacturing lead time from 36 days to 2 days, only to find that it took, on average, a further 24 days to deliver to his customer.

There are many recent advances in delivery systems. They are mostly vehicle related and fall in to three areas.

- Routing systems
- Scheduling systems
- Real-time communications and vehicle tracking.

Vehicle routing and scheduling systems have been used for several years and, with the dramatic reduction in the costs of personal computers, there are several commercial packages available in this area, producing good results.

Progressive systems in the area of vehicle management include the following (see Figure 11.9):

- The ability to allocate customer orders from a central scheduling facility to a number of depots according to the resources available at each depot, the volume of goods handled by the depots and the cost of direct delivery versus transhipment.
- Driver productivity systems which generate specific standards for the day or the route, that recognize the detailed handling requirements of orders and unique customer receiving conditions and methods.
- Customers are demanding higher levels of sophistication and flexibility within the base package, for example the ability to split loads for the same destination between vehicles or spread loads over several days. Other refinements include the ability to schedule collections and deliveries of the same goods on a single journey.

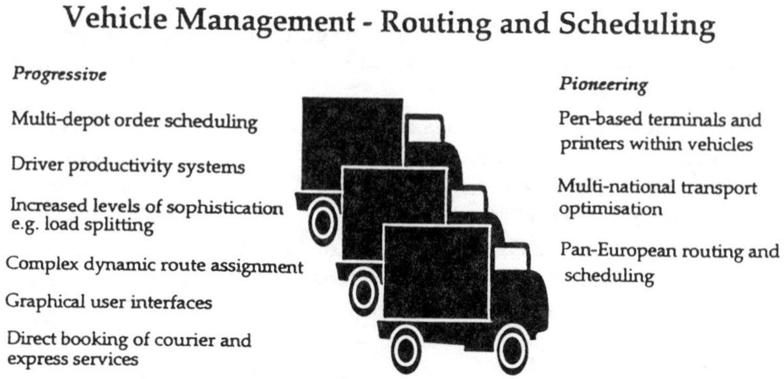

Vehicle Management - Routing and Scheduling

Progressive

Multi-depot order scheduling

Driver productivity systems

Increased levels of sophistication
e.g. load splitting

Complex dynamic route assignment

Graphical user interfaces

Direct booking of courier and
express services

Pioneering

Pen-based terminals and
printers within vehicles

Multi-national transport
optimisation

Pan-European routing and
scheduling

FIGURE 11.9. Vehicle management – routing and scheduling.

- Complex dynamic route assignment where the computer suggests routes and tasks into account detailed traffic and delivery patters. Integration with motoring organization information systems detailing road works and potential bottlenecks allows the route assignments to be more responsive to actual conditions. This is particularly useful as traffic congestion becomes more of a problem in the UK and the rest of Europe.
- The implementation of graphical user interfaces within routing and scheduling packages is increasingly widespread as more users demand this feature.
- Companies can now book courier and express transport services directly through computer terminals in their office. The terminals link directly to the contractor's despatching office for automatic booking and confirmation. Users also benefit from on line pricing information and instant delivery status reports.

Areas which are still pioneering for most organizations include the following:

- The use of pen-based terminals and printers mounted in the cabs of delivery vehicles to allow drivers to communicate with control systems and produce additional documentation, such as delivery notes and on-line POD (price on delivery) confirmation.
- Multinational transport optimism where there are several combined options for modes of transport (e.g. air, road, rail, sea). This is particularly useful when considering international transportation of goods and the rules that need to be set up to include desired lead time, minimizing cost and ease of handling, especially between modes of transport. Costs need to be tracked when the product is shipped across borders, since cross border shipping will always incur additional cost.
- Pan-European vehicle routing and scheduling requires map storage and display for all major European countries. However, this feature has not yet been effectively implemented.

Real-time communications and vehicle tracking are two quite distinct activities. The first is involved with communication between vehicles and their base station; the other is remote tracking of vehicles or automatic vehicle location. However, the real business benefits accrue from combining the two activities, since most users who track their vehicles require communication abilities in order to take advantage of the tracking information. Mobile communications are having a huge impact on logistics operations. There are several advantages of using this technology in fleet management and vehicle tracking: fewer delays, better management of truck movements on congested roads, and better allocation of slots for loading and unloading at warehouses. It therefore helps improve labour efficiency, fuel economy, and control over fleet management systems.

There are three main technology areas for mobile communications and tracking:

- Mobile Telephones
- Mobile radio communication (trunked mobile radio, private mobile radio, radio data networks)
- Satellite communications.

The choice of technology will be based on a number of parameters: cost (both initial and ongoing), system flexibility and compatibility, ergonomics, geographical coverage, upgradability and equipment design (particularly the robust and ease of use of on board terminals).

Mobile telephones provide the possibility of communication but a number of restrictions apply. Traditional systems only work in their country of origin. However, a pan-European mobile telephone network called GSM (Global System for Mobile Communications) is now emerging. GSM covers around thirteen European countries, but its coverage is limited to major cities and highways and will not service Eastern Europe. GSM will allow a driver from any participating country to use a mobile telephone in any other country. Mobile phones, unlike the other solutions, do not offer the ability to communicate data automatically and there is a low suitability for tracking purposes, since the location of the vehicle cannot be determined if the driver is present.

Mobile radio systems have much greater capability to transmit and receive voice, text or data messages. The major restriction of radio systems is that they can only be used for communication and tracking within the country of origin; however, as a single country based solution it is the most cost effective of the three.

The following areas can be regarded as progressive:

- A progressive technology in the mobile radio area is the use of public access mobile radio (PAMR), which is replacing conventional private mobile radio (PMR) systems. PAMR represents a cost effective alternative to both cellular and PMR communications. Fleet operators have traditionally had to bear the cost of buying PMR equipment (including handsets, transmission equipment and base stations). With PAMR, handsets can be leased or purchased from the network operator, who then provides network services and carries out maintenance. In addition, PAMR operators do not charge for air time since users only pay a monthly subscription fee for the service. PAMR, unlike mobile telephones, allows for mobile transmission of data. For example messages can be exchanged through a paging type system between a vehicle and its base station. Such messages may be transmitted to and from the on board terminal, and may range from short status messages (e.g. 'available for work', or 'unloading at Manchester') to long messages conveying detailed instructions, addresses and delivery schedules. Mobile radio also allows for the tracking of vehicles, although only for national operations.

- Currently, pan-European coverage requires the use of satellite systems for tracking and communication. In some countries (e.g. Eastern European countries) it is the only feasible method of communication. Satellite communication facilities are similar to mobile radio systems in that they allow the transmission of voice, data and text messages. Satellite tracking currently allows vehicles to be tracked by the base station to an accuracy of 300 meters. Vehicles can also be fitted with Intelligent Interface Units which are linked to on-board sensors and alarms to allow remote monitoring of vehicle parameters such as speed, direction and location. Back at the base station, the vehicle's position may be displayed on a PC screen, thus providing controllers with at a glance monitoring of all fleet movements.

- The combination of tracking and communication is regarded as progressive and has resulted in a number of key benefits, especially in the areas of customer service and increased efficiency:
  - A combined system can allow current positions to be relayed to base station for plotting on electronic maps.
  - Overall delivery time is reduced due to the ability to re-route a vehicle to avoid blockages, prepare customs to receive goods etc.
  - The vehicle's destination can be changed to a new collection or delivery point or can be relayed new information such as the need to collect two extra pallets.
  - Automatic status reports should be periodically provided to the base station throughout the day so that customers can be informed of delays. Cause of delays and exact location of vehicles.

- Load scheduling is improved. Thus, a driver can be sent on a time sensitive order without a back load, knowing that base station personnel have time to find a return customer before the driver has delivered his load.

- The collection and analysis of trip information by a removable memory card within the vehicle is regarded as progressive. When the driver returns at the end of the trip, the memory card is downloaded into the main computer at the base station. This information can be used for example, to

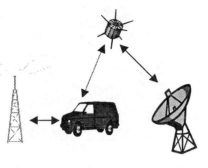

Progressive

Public access mobile redio

Satellite systems

Combination of
tracking and communication

Load scheduling

Collection and analysis of
trip information

Pioneering

Reservation systems for
road transport

Transmission of
load-related messages
e.g. temperature, barcodes

EDI and mobile
Communications

FIGURE 11.10. Vehicle management – mobile communications.

calculate the kilometres driven or without a load for the benefit of maintenance and invoicing. It can also be used to produce an immediate cost-benefit analysis per journey.

There are several pioneering practices with regard to tracking and mobile communications (see Figure 11.10):

- Combined mobile communication and tracking systems allow companies to have much greater visibly of their fleet. This has resulted in the pioneering practice of customers booking space on road transport in the same way that airline reservations can be booked.
- In addition to the storage of standard messages, load related information such as temperature and barcodes can be relayed back to control. The tracking of temperatures is highly critical for sensitive freight such as chilled food. An alarm is provided if the temperature on a vehicle strays beyond pre-set perimeters. This consists of a visual warning and an audible 'bleep' so that the staff at the control centre can contact the driver immediately with instructions about doing a repair or calling out help.
- A pioneering practice is the integration of mobile communication technology with EDI. This is covered later in the EDI section.

9   CUSTOMER ORDER PROCESSING

Although some distribution managers are not directly involved in order collection and order processing, every distribution manager is affected by changes in the methods used for processing orders.

Progressive order processing systems typically contain the following features (see Figure 11.11):

Progressive

Multiple methods
of order collection

Telesales and
telemarketing support

Dynamic matching of
supply and demand

Pioneering

Complex order
configuration checking

Artificial intelligence for
product selection

intercompany trading

Differentiated levels of
service

FIGURE 11.11. Customer order processing.

- The ability to handle multiple methods of collecting and entering orders into the computer, including hand held terminals, sales force PCs, telex, home shopping and EDI. Systems are now available that will accept orders from 'touch-tone' telephones and voice instructions. Although a number of these types of order entry are dealt with now, few of them are fully automated.
- Order entry systems which support telesales departments are becoming much more sophisticated and 'user friendly'. Extra functionalities, such as instant access to available stock figures, product look-up tables and customer specific item lists and price lists, are now included in a large number of order processing systems. These systems often also allow users to negotiate prices and inform customers of special promotion items. Information such as recent ordering patterns and payment patterns is available to sales persons while in direct contact with the customers. Other advanced features which support telemarketing activities include the use of scripts and call scheduling, providing 'active' customer targeting rather than 'passive' order capture.
- In line with moves to maintain service levels while reducing stock, many companies are investing in systems which allow allocation of scarce stock to take into account customer priorities. The system can either make a decision itself based on present algorithms, or present the options for adjudication by the customer service, marketing, stock planning and transport departments.

Pioneering features are:

- Order capture capabilities that include configuration checking for complex ordering combinations (e.g. for high technology equipment, related supply items and the customization of both prior to shipping).
- The use of artificial intelligence to assist with product selection and potential add-on items is emerging.

- Intercompany trading is a concept involving related companies within an enterprise. The most common definition is the ability for a sales company to take a customer order, pass the order on to a related company for delivery, with customer invoicing being followed by a settlement transaction between the two related companies. The ability to support intercompany trading on a pan-European basis through the use of EDI is described in the EDI section.
- The ability to differentiate between customers in the level of service offered is emerging as a key requirement for any customer focused organization, but as yet systems to support this concept are in pilot only.

10 DEMAND PLANNING

The areas considered above are all operational in nature, and deal with minute to minute activities. By contrast, demand forecasting is more tactical in nature. This area is important for a highly tuned operation because it affects service, stocking levels and ultimately logistics costs. It is a prerequisite for distribution requirements planning (DRP), mentioned below.

The most progressive systems integrate the sales, marketing, manufacturing and logistics activities into a forecasting process. This requires large amounts of data to be shared between the different activities and regular formal review and planning meeting to resolve conflicts and set priorities.

Such systems incorporate the following features (see Figure 11.12):

- Algorithms based on adaptive smoothing and employing warning mechanisms to indicate possible deviations.
- Forecasting at brand or product family as well as at item level.
- Aggregation of Forecasts to the company level prior to allocation at warehouse or regional levels.

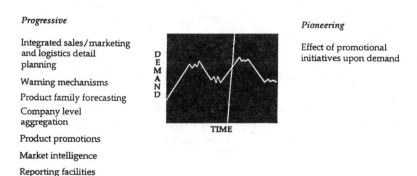

Progressive

Integrated sales/marketing
and logistics detail
planning

Warning mechanisms

Product family forecasting

Company level
aggregation

Product promotions

Market intelligence

Reporting facilities

Pioneering

Effect of promotional
initiatives upon demand

FIGURE 11.12. Demand planning.

- Handling new products, promotions, substitutions, suppressions and end of range situations.
- The ability to incorporate market intelligence information such as changes in product price marketing strategy and competitive information into the forecast.
- Reporting facilities to show errors in the various components of the forecast should be provided. Users should have the ability to select the desired level of aggregation together with the appropriate unit of measure. The ability to save forecasts on a logged basis so that forecast accuracy can be measured over a period of time should also be provided.

Pioneering demand forecasting systems have the ability to automatically recognize the effect upon demand of individual promotional initiatives. These more advanced features are usually difficult and expensive to implement, and require determination and significant resources. They are typically used for established branded products, but are also under experimentation on 'niche' or short life products.

11   THE ENABLING TECHNOLOGIES FOR RAPID RESPONSE MANUFACTURING LOGISTICS

Enabling technologies are the levers which will make possible significant advances in rapid response. There are three which are more significant than others. They are:

- Electronic Data Interchange (EDI)
- Client/Server Technology
- Expert Systems.

As with all technology one has to bear in mind that whilst the technology may have the ability to make things possible, it is people who ultimately make things happen.

12   ELECTRONIC DATA INTERCHANGE (EDI)

EDI has generated considerable excitement and confusion in the European business community. EDI or 'paperless trading', is the computer to computer exchange of information between trading partners using agreed upon standards to structure data. The transmission is achieved through public switched networks, public or third party networks, or private leased lines for large volumes of data. Communications may be point to point or point to multipoint.

EDI can mean different things to different people. Electronic mail, which is now fairly commonplace, and other simple model based file transfers are forms of EDI because they transmit data electronically from one computer to another. Companies which seek to integrate information systems with their customers and suppliers through EDI are focusing on two key challenges: the selection of appropriate message standards and the establishment of trading links.

EDI offers a number of benefits at both operational and strategic levels. At the operational level, existing business processes can be improved or reengineered. The most tangible benefit is in the more efficient transfer of business documents. EDI helps reduce the amount of paperwork required, the number of errors associated with manual transactions and the incremental cost of a transaction. Within the EC alone, it has been estimated that the cost of paper used can be reduced by ECU 5.9 bn through the use of EDI.

By reducing manual, error prone handling of documents, EDI can also reduce the number of exceptions to be dealt with and the associated administrative costs.

Increased speed of communication and greater real time interaction between business partners can have a significant impact on logistics activities. For example, EDI enables shorter lead times, and allows more accurate and frequent ordering which are essential for the effective implementation of just-in-time operations. This enables the reduction of stock levels and leads to fewer stock outs. Moreover, if order lead time can be reduced sufficiently and the supply is reliable, stock holding can be completely eliminated.

This has led directly to the development of cross docking, which aims to move product from the supplier through the distributor's distribution system without putting the product into storage, and is discussed in greater detail in the cross docking section. Although such a strategy incurs risks of supply chain failures which could lead to severe marketplace impacts, EDI based sharing of information with customers and suppliers may offer different parties advanced warning of each other's intent. For example, suppliers no longer have to second guess their customers' requirements since sales forecasts, stock levels, delivery schedules and advanced shipping notices can be used by manufacturers to help them best plan their production, and by customers to best plan their product intake and to take advantage of price discounts available through extended production runs.

EDI must be seen as a co-operative activity, since if used properly it can be used to build and strengthen long term strategic relationships between customers and suppliers in areas such as joint product development and logistics partnerships. An example of this is the development of ECR (Efficient Customer Response) which, through the use of EDI, seeks to improve, for example, the grocery supply chain, minimizing inventory levels and optimizing product availability and product quality. The concept relies on partnerships being formed between manufacturers and retailers who together optimize stock levels. Thus, a retail promotion would involve regular stock replenishment to meet consumer demand rather than stockpiling by either manufacturer or retailer.

Successful exploitation of EDI technology to gain competitive advantage requires that two or more organizations enter into its implementation with a shared view of how best to meet a customers needs. However, it must be appreciated that there are significant barriers to EDI implementation. The disadvantages

include the need to invest in internal systems, the need to develop customer interfaces, and the length of time to establish trading links. These obstacles can be
particularly difficult for a company's first installation because few firms have
internal interfaces required on their host systems to accommodate EDI. Additional
possible disadvantages lie in the lack of knowledge and confidence to stimulate
their trading partners into action, and in the need for most companies to acquire
in house functional and technical expertise in order to champion and deliver the
trading links.

The following are regarded as progressive within EDI (see Figure 11.13):

- If requested goods are not available for a customer order, it has long been
  possible to allow a system to automatically propose a product to substitute
  it. What is pioneering is to communicate the substitute back to the customer
  and allow them to decide whether to accept it, before proceeding with the
  picking. This closed loop mechanism, implemented with EDI, overcomes
  many difficulties often associated with substitutions.
- With many companies now using third party carriers to distribute their
  goods, the use of EDI links with carriers can lead to quicker responses and
  higher efficiency. When supplied via EDI with load and destination details
  in advance, carriers can perform more efficient route planning. If carriers
  return proof of delivery notes by EDI, this enables quicker reconciliation
  with orders and facilities more rapid payment.

Progressive

Product substitution

Third party carriers and EDI

Transmission of sensitive information
e.g. price data

Integration of EDI into business
Application e.g. warehousing,
Manufacturing, stock control

Transmission of supplier data e.g
Sales acknowledgements, automatic
Shipping notices, etc.

Pioneering

Integration of EDI with mobile
Communications

Adoption of EDIFACT standard

Cross-border use of EDI

Electronic mail within EDI

Transmission of EPOS sales dat
to supplier

Intercompany trading networks

Broadband / ISDN

FIGURE 11.13. Electronic data interchange.

- Although EDI is widely used for order processing, invoicing and payments its use is being extended to encompass the data areas. For example, some large retailing organizations have implemented EDI still experience problems with around 10% of their invoiced lines due to differences between agreed prices and invoiced prices. Thus, sensitive information such as price data is now being transmitted through EDI, ensuring that any queries will be resolved earlier permitting a better pass rate for the invoices.
- To reap high benefit levels from EDI, it must be integrated into the company's other business applications for warehousing, receiving, manufacturing etc.. Some businesses use EDI to receive purchase orders and then print them for subsequent use in manual procedures. This situation clearly does not harness the full power of EDI and, arguably, similar benefit levels would be achieved through the use of facsimile technology. An example of an integrated EDI transaction would be the following: an order processing system receives EDI purchase order data and passes it directly to the next application, which accesses an on-line price catalogue, checks the database to verify the customers credit and then passes the order to the warehouse. A barrier to the above situation is that each EDI implementation will need to be customized to match the application architecture and current workflows of the system. Increasingly EDI vendors are compiling lists of the application software packages that they interface with, but it is unlikely that these will be complete 'off the shelf' plug and play systems.
- Suppliers are increasingly sending a variety of documents using EDI, such as sales acknowledgements, picking lists, automatic shipping notices and invoices on shipment of a product. This enables the buyer to plan inbound logistics and allocation of product.

The following areas are regarded as pioneering with EDI:

- The integration of EDI with the mobile data communications (MDC) arena. Standard messages are currently being developed within EDIFACT standards for use within EDI applications. Although there is some overlap between EDI and MDC (e.g. status reporting from vehicle), in general EDI messages have different contents from those of the MDC area. The EDI messages are much more structured than those of MDC, since they conform to internationally agreed formats. The EDI messages contain information about transport orders, preliminary and final bookings, the contract and the status of the transport (e.g. condition and temperature of cargo). The MDC messages contain loading and unloading instructions, confirmation, expected time of arrival/departure etc.. This integration will enable organizations which currently use EDI for commercial data exchange to use essentially the same software in order to find out where their vehicles are and to receive information about traffic congestion or bad weather conditions. An example of EDI and MDC integration is the information link between consignor and shipper. When a driver has delivered a consignment, a standardized message is sent to the base station via mobile communication, stating that a consignment has been unloaded. At the base station, this message is received by the in house computer system which automatically translates the message to standardized message for the consignor and message for the planner. This has a number of key advantages. The number of key advantages. The consignor and the planner immediately know that the cargo has been delivered and this information can be used in further planning activities. The consignor does not have to be informed directly by the planner since this has already been achieved by the system and leads to a reduction in the planner's workload.

- Standardization is essential to EDI success. Transactions prepared in an in house format need only be converted into one standard format in order to facilitate electronic trading with many other partners. Furthermore, the use of these standards helps prevent technological obsolescence. Today, EDIFACT represents the major international, cross industry, general business standard. There are a number of industry specific standards (see Table 11.2). Through further development, these standards are starting to converge on EDIFACT as the common message set. However, this process is far from complete and is a major barrier for companies wishing to implement EDI on a pan-European basis.

- Initial EDI implementations have been nationally focused. One of the reasons for this, at least on an European level, has been the differing EDI standards. Other reasons are the weakness in the pan-European telecommunications infrastructure and a lack of globally organized network operators and service providers. However, some retailers are attempting EDI implementation on a pan-European basis. Initially, the areas that have been successful nationally, such as orders and invoices, are being implemented internationally. In the longer term, other documents for shipping and customs will be brought into the EDI framework.

TABLE 11.2.

Standard	Application area
TRADACOMS	RETAIL
ODETTE/AIAG	AUTOMOTIVE
EIDX/EDIFICE	ELECTRONICS
CALS	MANUFACTURING (MATERIALS/PART/SPEC)
RINET	INSURANCE
SWIFT	BANKING
EANCOM	RETAIL AND COMMERCE
CEFIC	CHEMICALS

- Some pioneering companies are also starting to introduce electronic mail and this offers particular benefits when used with international suppliers. First, there is none of the delay inherent in posting a document since transmission and receipt are almost instantaneous. Electronic mail is more efficient than using the telephone, especially if there is a difference in time zones. For example, Hong Kong is eight hours ahead of the UK, giving a very small time window during which it is possible to speak with a person. Electronic mail maintains the personal contact but with clear, concise and convenient messages.
- Sales information captures at the point of sale by EPOS systems is transmitted to the supplier. However, it should be ensured that this information is limited to the supplier's products and that any competitor data is eliminated. This data helps the supplier to plan resources and capacity need to meet customer requirements. Through the use of this technology, it should be possible for suppliers to manage replenishment of selected products.
- Another pioneering feature is the ability to conduct intercompany trading on a pan-European basis using EDI. The system should be able to indicate the cost tax efficient method of executing the intercompany trading, while the EDI link ensures that much of the cumbersome administration associated with intercompany trading is reduced.
- A significant increase of ultra high speed networks (broadband/ISDN). This will also result in a significant opportunity to improve on location and shipment tracking accuracy.

13   CLIENT/SERVER TECHNOLOGY

Client/Server technology can be defined as the splitting of an application into tasks which are usually performed on separate computers, one of which is a programmable workstation (e.g. PCs see Figure 11.14).

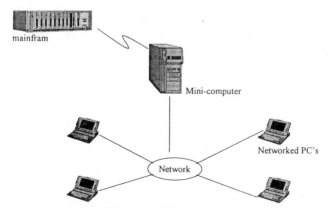

FIGURE 11.14. Client/Server architecture.

There are some key drivers in the trend towards client/server technology within the logistics software area. Organizational streamlining is increasingly being pursued to cut overall costs (i.e. by reducing the number of bureaucratic, non-value-added positions), to achieve faster responsiveness by authorizing local decision making, and to tap into greater innovation by utilizing local expert knowledge. This creates major new demands on cross functional information access delivery. Organizations should be provided with the tools and training for users to be able to operate on their own.

Another factor is the continuing desktop price/performance improvements. In 1980, the cost per MIPS (million instructions per second) of mainframes and minicomputers was 15 times greater than our workstations. In 1990, it was 10 times greater and by the year 2000 the cost could be anywhere from 700 to 2000 times greater. It is now possible for workstations to have processing power equivalent to the fastest IBM ES/9000 mainframe processor.

Although many users do have concerns about technology and vendor reliability, it should be appreciated that client/server technology enables users to obtain cost effective power from workstations. Another key driver is the workstation's capability to enhance substantially the user interface. Traditional host based systems are technically bound to form filling interface in which users are prompted via textual hints and instructions to communicate with a system by filling in fields on a terminal screen. The complexity of field and form interrelationships and the limitations of textual communication creates a need for specialized training around each application. For example, one set of employees are trained in order entry, while others are trained in customer service.

The implementation of client/server technology enables the use of applications which use not only text, but audio, image and video communication as well. A high degree of interactively permits the user greater control over the flow of the dialogue and allows the system to only solicit information which is relevant to current and previous inputs. Another important driver is to communicate with a number of distinct local and remote processing systems. From a single workstation, a user can access vast amounts of processing power and associated data, but without being aware of the location of the data or the machine.

The above drivers of client/server systems have a large impact on logistics software packages. Complex areas such as warehouse simulation, computer aided picking and supply chain modelling can now be implemented much more efficiently. Mainframes can still be used as powerful database servers, while the workstations handle the majority of the processing and graphical user interface functionality. Client/server systems are also potentially suitable for pan-European logistics management, since this requires both central and location specific processing. Central processing is required for identifying and monitoring macro-level information such as market demand, stock levels and current production plans in order to make the best decisions for distributing goods on a European basis. At the same time, micro-level processing for order entry, warehouse management, delivery management, invoicing and accounting is essential. Thus, client/server technology enables organizations to operate more efficiently without imposing a particular structure, since it is possible for organizations to be centralized or de-centralized, product or geographically oriented.

Client/server systems are implemented on a number of different hardware and software platforms such as Unix, Windows/NT and OS/2. Unix is increasingly favoured as a software platform due to its maturity in relation to the other two platforms and its scalability and portability. Scalability allows companies to implement different capacity computers at different sites. For example, a regional warehouse may only require a five user computer system whereas the headquarters system may need to support 50 users. Portability allows a Unix system to be implemented on a variety of hardware architectures. However, the portability of Unix software has to be qualified since two competing organizations, OSF and Unix International, are responsible for the development of core Unix routines. The entrenchment of these two camps means that the existence of a single Unix standard could take many years to emerge.

Market leaders are increasingly migrating their packages to a client/server architecture on a Unix platform. The primary effort has be focused on migrating the packages to Unix, rather than developing sophisticated graphical user interfaces (GUI) or remote data management access strategies. These packages are not as strong, in a functional sense, as their counterparts on proprietary hardware.

Client/server technology can be viewed as a key driver in business reengineering since it increases the flexibility and effectiveness of logistics planning and management at both the macro levels. Finally, it should be emphasized that the implementation of client/server systems is a complex arena. This is partly because development and maintenance of these systems is more complicated because the incorporation of event driven GUIs requires more intricate logic than traditional character based systems. The extensive communication requirements of these systems also significantly increases their complexity.

14   EXPERT SYSTEMS

As we enter the knowledge age it is important to recognize the role of expert systems in the more general subject of artificial intelligence. Artificial intelligence (AI) is beyond the scope of this chapter in that only very few companies will even consider themselves to be on the edge of that aspect of AI which implies the use of an inference engine in which the computer could start to make decisions. This is certainly not a trend. Indeed, Bill Gates writing about reality of the virtual office in Andersen Consultings' *'Outlook'* magazine prefers to refer to IA – Intelligence Augmented, *'Take my intelligence and augment it. Don't try to replace the individual'*.

This is the view taken by George Hess at Ingersoll Milling. He says of expert systems:

> 'The reason that world class manufacturers are so excited about the use of expert systems is simply because of the opportunity that it gives them to increase productivity. American and European industry is really just

beginning to awaken to the benefits of expert systems, while the Japanese are now moving into their eleventh year of concentrated artificial intelligence research and applications. This is part of the 'fifth generation' computer project that was started in April of 1982 by coalition of Japanese manufacturers in co-operation with their government. It is predicted that these knowledge base systems that are being applied both in their design and manufacture are expected to help the Japanese dominate the markets for their products. With the productivity gain opportunities that we see in expert systems, it is obvious that the survivors will be using them extensively by the mid 1990s.

Since expert systems are so widely applicable, so promising, so powerful and so near to being production worthy, they merit an immediate pilot effort in every organization. These will not replace the integrated systems, but they will be incorporated directly as part of them. Expert systems, or more properly knowledge based systems, will be used in place of the heavy logic of present day procedural systems. They will capture the logic of our very best experts in all parts of the business, and make that logic available to the less experienced person operating the system. In this way, everyone will be operating at a new and higher level of performance. The expert over a particular 'domain' of the business will still be improving his or her logic and upgrading the systems knowledge base. The operating novice will use this system to function approaching (not attaining) the performance level of the best expert in the business, thus improving both positions, but most of all insuring that the business operates nearer to the optimum level.

A key future of the expert system is its non-procedural nature. This allows prototype systems to grow into fully fledged expert systems (ES) without the expensive reprogramming that we experience today. When a prototype ES needs to be improved, you add more rules and regenerate the knowledge base, thus giving the inference engine more knowledge to work with, but you do not have to redesign the system.

One thing that you must remember is that no matter how good you are in designing an expert system, it will not perform as well as your best expert. The better the knowledge engineers are in pulling out and recording the knowledge, and the better the experts are in expressing their knowledge, the closer the system will approach the expert's performance. At our present stage of the development, the best we can do is assist the expert and attempt to replicate his or her performance for use by the lesser experienced people in the organization.

You should not try to have the expert tell you what he or she knows. You should concentrate on what they do, because after all that is what you are trying to replicate. It is also much less threatening to the domain expert if you take that approach, because the domain expert probably does not know what he or she knows, but certainly does know what he or she does.

The real opportunity for use of expert systems is not as new super-human systems but rather intelligent enhancements to augment conventional systems'.

So between Bill Gates and George Hess, representing two companies leading technology in different fields we see the trend towards expert systems in their use to augment what already exists.

A parallel trend in this field is towards rules based systems which focus on capturing the method or rules by which your best expert arrives at his conclusion rather than capturing the conclusion itself.

Major aerospace companies like Boeing and General Electric appear to be moving towards those technologies which are rules based and provide a seamless continuum of technology from initial design concept to find customer service programme after delivery. In so doing they are infinitely better able to predict in advance of major capital investment the lifetime cost of components and aircraft.

This trend is supported by leading edge technologies from emerging companies like Cimplex Corporation whose knowledge based computer aided manufacturing system from feature based modelling finally automates programming of machine tools and other equipment in the manufacturing process.

In applying these advanced technologies it is worth noting the advice in William Taylor's book *'Demystifying Artificial Intelligence'* (Graeme Publishing Corporation, 1986) in which he advocates the trend should be towards conservatism:

'Concentrate in areas in which apprentices are routinely trained – if experts are accustomed to training amateurs, they will be more patient with computers.

Make sure the problem can be solved by humans – if there are no experts, expert system technology is inapplicable. Expertise need not be rare for an expert system to be profitable, but there should be a high value on having more experts.

Avoid problems that cannot be quantified – cotton graders feel a fabric sample and agree on the grade, but no one knows what they measure.

Look for a high payoff application – developing expert systems is costly and time consuming, so a problem that is worth solving could be chosen.

Make sure users want it – an expert system that works will be of little value if it is not used. Find a way to make it easy and desirable for the users to use the system even if possible, make it a part of their daily routine.

Do not risk anyone's career – the odds are that the first expert system project may not pay off as a measured by an ROI, but it will undoubtedly be a very valuable learning experience for those involved. Unless the investors are tolerant of a few false start, it is unwise to begin.'

Further trends in thinking of the leading companies in expert systems are embodied in a SME (Society of Manufacturing Engineers) Blue Book entitled *'Expert Systems – How To Get Started'* (1/800/733-4SME). It was written by a team led by Ingersoll milling but incorporating General Electric, General Motors and AT&T.

15 CONCLUSION

Trends by definition are timeless in their adoption. Without doubt the key constraint to the adoption of Rapid Response Manufacturing Logistics is people, their attitudes, their experience, their cultures and their willingness to change. Environmental issues will also have their roles as demands from environmental legislation the resources of innovation from other areas.

In Europe there is undoubtedly a renewed interest in time based performance measures which is speeding up supply chains. There is a new strong interest in EDI and logistics conferences remain well attended. Customer service is on the lips of most responsible managers and reasonably good understanding exists of what this means.

Whether there is enough focus and determination to achieve world class efficiency is impossible to judge. In The United States things change more rapidly than Europe once they are under way. The recent established Rapid Response Manufacturing Programme is a good example of the starting situation.

The Rapid Response manufacturing (RRM) programme is a five year collaborative effort involving General Motors, Ford, UTC/Pratt & Whitney and Texas Instruments as end users along with five advanced software suppliers (Aries, Cimflex Teknowledged, CIMPLEX, ICAD and Spatial) focused on supporting automated concurrent engineering. Total funding for the programme is set at $45.8 M. $26.8 M is provided by the industrial partners and $19.8 M is provided by NIST through Advanced Technology Programme (ATP). Through the generic Cooperative Research and Development Agreement (CRADA) recently signed with the DOE, the Oak Ridge National Laboratory (Martin Marietta Energy Systems) is providing an additional $4.5 M in resources for technology testbeds. The RR Cooperative Agreement with NIST began October, 1992.

The goal of the programme is to reduce the product to market time by 50% through the application of advanced computer technology to leverage concurrent engineering methology. Key objectives in support of the above goal include:

- One pass product design
- Direct access to all product design information throughout the life cycle
- Knowledge assisted trade-off analysis
- Manufacture an accurate first part demonstrate interoperability of systems architecture
- Create and use variant design processes.

Important guiding principles include pursuing open architecture that can be shared by suppliers. The intent is to implement available technology by achieving interoperability between software applications and across existing legacy systems used by the partners through use of an open architecture.

This type of initiative is essential to the speedy advancement and proliference of rapid response manufacturing techniques and indicates that the USA is probably on its way to the next generation of exceptional performance. In Japan, as indicated at the start of this chapter the performance gap is probably still widening.

All of which indicates that there is no respite for European manufacturers who want to remain in the global game. Ultimately each company must decide whether the adoption of these trends in rapid response is right for it. If the answer is positive, there is no time to be lost.

ACKNOWLEDGEMENT

The author is grateful to his many colleagues at Andersen Consulting who have contributed to the preparation of this chapter, several of whom prepared the text for the sections of their expertise.

REFERENCES

Andersen Consulting (1994) Logistics Software Guide.

Copacino, W. C., Ernst, K. R. and Richmond, B. S. Quick Response: Is it right for your company? *Logistics Perspectives*, Andersen Consulting.

Hess, G. The SME New Manufacturing Enterprise Wheel: A real world example in operation today, Ingersoll Milling Machine Company.

Wrennall, B. Handbook of Commercial & Industrial Facilities Management, ISBN 0-07-071935-7.

12 Quick Response Manufacturing in the Tyre Industry

P. A. Folwell[+] and C. I. Jones[*]

[+]HOSCA Management Consultants Ltd.
[*]SP Tyres UK Ltd.

1 INTRODUCTION

Having spent several years in automotive component supply, and in particular the tyre industry, the authors of this chapter have witnessed at first hand the radical changes which have hit the industry in recent years. This chapter will examine the background and historical context of the growth of manufacturing which has led to its present position. It will then seek to describe some of the relevant practices and theoretical frameworks prevalent in the 1990s; in particular, the techniques that are used within manufacturing. This will then be followed by a more detailed description of the tyre manufacturing process, its components and raw materials and their suppliers. Finally, the authors will demonstrate how different parts of the tyre making process are more adaptable to quick response manufacturing than others.

2 HISTORICAL SETTING

Since the theme of this paper relates to the supply of tyres to the automotive industry, it is inevitable that many of the references in it are to that sector of manufacturing. The most influential research conducted in this field has been by Womack et al. (1986) with their massive $5 million programme which culminated in the publication of 'The Machine that Changed the World'. Other contributors not only to that research, but also with significant work in their own right, would include Richard Lamming.

The system of mass production that exists today had its origins in the latter part of the nineteenth century in the United States. Before that time goods were manufactured by the craft system whereby each product was bespoke, and therefore unique. Even products that appeared to be the same were in fact individually assembled by careful fitting of the components by skilled craftsmen. At the same time, the way in which companies were organized was very different to today. Many of the major American corporations used a system of 'Gang Bosses' who were, in effect, sub-contractors who had the responsibility of hiring and firing entire crews to work in factories (Montgomery, 1979). The power of these gang bosses was virtually absolute, since they could determine not only who worked in the factory, but also his rate of pay. Within certain communities the gang bosses wielded enormous power and were often 'bribed' by people to secure jobs for them. It was within this environment that Frederick Winslow Taylor developed not only his own career, but more importantly the basis of his Scientific Management theories. Taylor must have seemed like the answer to American industry's prayers; by adopting scientific management techniques they were able to regain control of their own workforce and make seemingly incredible improvement in their productivity.

The move from a craft-based organization to mass or flow-line production can be illustrated by the fact that each worker at Ford's plant had an average task cycle time of 514 minutes in 1908 which had been dramatically cut to 2.3 minutes by 1913 (Womack *et al.*, 1990, pp. 28). Another example is how Henry Ford adopted Taylor's principles at his Highland Park plant in about 1915. It is worth noting here that Taylor's major work, 'The Principles of Scientific Management' was first published in 1911 and that whilst Henry Ford may not have actually read the book, he would certainly have been aware of its content. When the plant was originally opened it took an average of 12.5 hours to build one car and Henry vowed that he would produce one every minute. Within 12 years he did exactly that and within another 5 years had reduced it still further to only 10 seconds.

The other strategy which had a major effect on Ford's domination of the market and on the beginnings of mass production was the instigation of standards coupled with a ruthless supplier relationship and eventual vertical integration. With the exception of companies like Eli Witney in the early 1800s, who had already developed a form of component standardization for the manufacture of large quantities of firearms (Bessant, 1991, pp. 18), the practice was by no means widespread. In the late nineteenth and early twentieth centuries the usual routine was for the newly emerging car manufacturers (often general engineers themselves) to sub-contract the component manufacture to smaller engineering workshops. Standard gauges were not widely used and therefore the final assembler at the car factory was forced to 'fit' these somewhat approximate components together. The problem was that components were made on 'soft' tools which would tend to wear very quickly as they were used. The greater the wear, the larger (or smaller) the components became; so that a complete range of sizes from ostensibly the same set of tooling was perfectly normal. The advent of steels that could be hardened meant that Henry Ford was able to insist that specific tolerances were adhered to by his sub-contractors; thus eliminating the need for final fitting and enabling incredible advances to be made in cycle times.

So, hardened steels led to standard components which in turn led to the elimination of 'fitting' during final assembly and resulted in previously unimagined cycle times. It would therefore be extremely unrealistic for anyone to have commented to Ford at the time that his stocks of semi-finished goods had risen as a result. He would simply have laughed and pointed to the fact that his unit costs had shrunk and that these had been passed on to the customer, which in turn had created a massive demand for his products.

The other significant player in the motor industry at this time was Alfred P. Sloan who took over General Motors from the founder William Crapo Durant in 1921. Durant had been extremely successful at expanding his company via acquisition; notably Buick, Cadillac and Oldsmobile, but had totally failed to integrate them into one company and was eventually forced out by the banks (Halberstam, 1986). Sloan did two things which, when added to the start that Ford had made, effectively completed the change to mass production. Rather than attempt to integrate the subsidiaries into one large organization, he decentralized but maintained rigid control with a superbly detailed reporting system, and at a time when Ford was offering only the Model T ('any colour you want as long as it's black'), Sloan developed a broad range of products to suit what he perceived to be an equally broad range of customers.

Manufacturing has thus continued along the path started by Taylor and developed by Ford and Sloan right up to the present day. In fact, the vestiges of Taylorism are still rampant in many so-called modern companies and manifest themselves in the autocratic management styles and in the 'incentive' schemes which they operate.

3   MANUFACTURING TECHNIQUES

In recent years the complacent attitudes of Western businesses have been shattered by the onslaught of Japanese manufacturing methods which have finally forced the West to re-examine its own business philosophies and organizational systems.

After World War II, Japan was forced to enlist the help of the Americans in rebuilding its industrial base. Among the people who visited Japan at that time were J. Edwards Deming and Joseph Juran who had tried, unsuccessfully, to promote their theories in the States before the war. It was Deming, Juran and others like them who were largely responsible for inculcating a quality ethos in Japan that has to a great extent been the reason for their undoubted success. It is interesting to note that when the West eventually realized that they were losing their market share, they turned to Japan in an attempt to learn their secrets, only to be told to look closer to home.

Japan is a relatively small country with little or no natural resources of its own. In the years following the war their domestic market was also small and diverse. To prevent a complete take-over of Japanese industry, the Ministry of Trade and Industry – MITI stopped foreign investment and imposed a 40% import tariff on all foreign made vehicles (Lamming, 1993, pp. 17). The effect of this was to strengthen the position of the domestic market and to allow the indigenous motor industry time to develop.

The Japanese motor industry was certainly not new, and firms like Toyota had been producing vehicles since the 1930s. But Japan could never hope to compete against firms like Ford and General Motors on their own terms since they could not match the massive economies of scale enjoyed by those companies. Taiichi Ohno, head of Production Engineering at Toyota, worked closely with his chairman Eiji Toyoda to develop a system which was better matched to their capabilities, capacities and market. This is was what was eventually to become **Lean Production**, the term coined by Womack *et al.* (1986) to describe this form of manufacturing.

Of the many techniques which are being increasingly adopted by Western industry probably the most relevant to quick response manufacturing are:

- Simultaneous engineering.
- Supplier collaboration.
- Cell manufacturing.
- Set-up time reduction.
- Just-in-time.
- Kanban.
- Kaizen.

It is important to realize that the concept of lean production applies not only to the supply and delivery side of a business, but also to the process of product development and launch. By means of such activities as Simultaneous Engineering it has been possible to drastically reduce the time taken to bring new products to the marketplace. This in itself has to be one of the most important advantages any company can have over its competitors.

Another extremely important aspect of quick response manufacturing is for companies to develop closer relationships with both their customers and suppliers. The complete supply chain needs to be carefully constructed to work in harmony and to avoid the traditional confrontational relationships that previously tended to exist. Whereas in the past it was normal practice for a company to have as many suppliers as possible on its books, it is now far more usual for major companies to co-operate with a much reduced number of sub-contactors or partners than before.

One illustration of this is the increasing tendency for car manufacturers to effectively delegate more design and manufacturing responsibility to their suppliers. The extent to which this arises can be seen by the proportion of the design content undertaken by the car manufacturers themselves. Whilst in the United States 81% of the work is in-house, in Europe this figure drops dramatically to only 54%. However, these figures must then be compared with the level in Japan of just 30% (Clark and Fujimoto, 1987; 1988a; 1988b; 1989). It is this reliance on the supplier which many car manufacturers consider when making their choice of partners. It is far easier to be able to openly discuss design requirements and having product designers based locally can provide the supplier with a competitive advantage over those organizations whose design teams are perhaps located in an overseas headquarters.

Cell manufacturing, Just-in-time manufacture and the pursuit of reductions in the time it takes to effect a product or size-change are all integral components of a company's quick response strategy. The use of cell manufacturing enables a degree of flexibility on the shopfloor which is more readily able to accept sudden changes in demand. The installation of relatively small, and perhaps slightly less technologically superior machines make it possible for operators to communicate with one another and thereby resolve their 'problems' together. If one of the processes in a sequence suffers a breakdown, there is little point in continuing to push production towards it. Far better to stop producing, fix the fault together and then resume.

Many companies have decided in recent years to adopt a policy of Just-in-Time manufacturing without really knowing precisely where to start. Perhaps a more practical method is to create test cells and then train the operators in problem solving techniques. From this can come the vital reduction in the time it takes to change from one size to another and with the help of a Kanban system, just-in-time production would almost certainly be the result.

Finally in this section it should be stressed that without the full and willing co-operation and participation of the workforce none of the above would be possible. It is therefore essential that not only the major car manufacturers but also the smaller component suppliers use all their available resources and not rely solely on the management. Companies should endeavour to introduce such participation schemes as Kaizen (Imai, 1986) which aims at involving the maximum possible numbers of employees in working for continuous improvement. At SP Tyres, the organization that was inherited in 1985 consisted of the traditional management/workforce demarcations which had existed for years. To become a leading manufacturer in a highly competitive industry this situation could not continue – there had to be a radical change. Much of that change has been possible through the use of team working and in particular, Kaizen.

4   TYRE MANUFACTURING FLEXIBILITY

The need for an increasing diversity of products, with consequent needs for flexibility are evident to all today. The pressure to drive costs down has forced a wider view to be taken of the manufacturing/supply process. Whereas each production area would traditionally be targeted to improve the costs within their own shop, without reference to the whole process, it is now normal to look at the whole process. Improvements to the whole will certainly mean that costs will rise in some areas, offset by far greater gains elsewhere in the production/supply chain.

The change in outlook in the tyre industry has come about in the last decade. One major tyre manufacturer was expanding in Europe in the early 1970s with several near identical car tyre plants set up to make just four sizes each. Now these plants will be running 8 times that number simultaneously as a result of the increased variety demanded by the customer. Although one significant tyre producer continues to espouse the benefits of concentrating sizes in different factories throughout Europe, they may be attempting to make a virtue from a vice, i.e. they may well have low flexibility within their individual plants.

An alternative approach is one that is based on profiting from the flexibility at each plant, to enable the widest range to be produced as close as possible to the customer. Support, on a European basis, for severe fluctuations in demand in any particular market is more readily accomplished through the flexibility of the whole group.

5   THE TYRE-MAKING PROCESS

5.1   Curing

Improvements in flexibility have generally come about in reverse order to the manufacturing process. After curing, multi-diameter chuck uniformity machines are the norm.

Typical SMED (Shigeo Shingo, 1985) studies have been carried out in tyre curing shops. To reduce the temperature problems, cooling the mould by a fine spray of water on it and preheating the replacement have contributed to dramatic time reductions in what is classically the bottleneck shop in a tyre plant. Having removed this obstacle, the next area that has been tackled (this time in many cases by high capital investment) has been in the tyre building area. This has not been confined to car tyres, but has also taken in truck tyre plants.

5.2   Tyre building

Tyre building equipment evolved in the 'normal' way, with effort concentrated on improving output and hence productivity and on improving quality by reducing the manual skills needed. Daily outputs increased and whereas the old machines could be matched against two moulds, the new equipment produced enough for four moulds then six moulds, with further evolution possible to eight moulds. One consequence of this was to dramatically increase the number of moulds purchased and another to increase the stockholding of tyres to an unsupportable level.

This evolution was happening at a time when the range of tyres was increasing, by specialization. Instead of an all round tyre, new patterns/constructions were being introduced for drive axle, steering axle and trailers, along with on-off road tyres, to better satisfy users, whilst demand continued for the all round patterns. This growth in variety accelerated the realization that further productivity gains in the tyre building area would impose a disproportionate increase in costs elsewhere and forced a rethink.

The search for flexibility has involved one leading tyre manufacturer in supporting four competing in house multi dimensional tyre building machines, each with a capital cost of up to $3 million and a corresponding resource cost in both hardware and software design. Part of this cost was offset by the reduced numbers of curing moulds that would be required and in reduced stockholding needs that are entailed with production of large batches infrequently.

One other problem of these increasingly large machines, (with control cabinets taking up more space on their own than the equipment replaced), is that of isolation of these operators from the rest of the factory. Whilst it was recognized at the outset that they need to work together as a team on the machine, teamwork across the workshop is not helped by the layout of the machines.

For car tyre building machines, engineering solutions have also been sought, in order to accomplish a wide range of sizes, with good productivity. The culture of a company can have an important bearing on how this is achieved. If technicians are excluded from statistics from which decisions are made, savings in the cost of production operatives have in some cases been more than offset by a rise in the number of highly qualified machine control software technicians employed. The time required for the introduction of such equipment can also become dependent not on how long it takes for the mechanical design and construction, but on how soon the man years of machine control software can be completed.

5.3   Extrusion

Increasing the number of sizes being made at any one point in time can have a dramatic effect on the next stage upstream. Traditionally there has been productivity pressure on extruders, for ever increasing outputs, normally specified in kg/minute, at ever improving consistency. Now, to this pressure has been added the need for the ability to cope with a wider range of compounds and a larger number of components. Thus rapid change ability, creating the minimum waste materials, is now an essential part of a modern installation.

The bottleneck in some tyre manufacturing plants has been transferred upstream, from curing to extrusion, by the explosion in component numbers from running more sizes concurrently.

At this point it is necessary to highlight an important difference to the metal working industry. The pressure to minimize stocks of work in progress has been with the tyre industry since it began. Whilst steel components will remain usable if kept dry, or protected, this is not the case with rubber components. These each have a relatively short usage life, with normally a minimum time after processing to allow the product to cool and stabilize and a maximum time after which the components are too dry or stiff to use. Wasteful scrap or rework result from not getting the quantities right. Kanban systems work well in these areas, which depend on the reliability of the equipment and the rapid feedback of requirements, from 'customer' to 'supplier' and *vice versa*.

Responding to ever more exacting and differing needs by OE (original equipment) customers, for tyres exactly in tune with their car, can also result in a proliferation of components into what looks to the layman to be an identical product. Whilst the moulds may be unchanged, the carcass going into those moulds may vary. This may not be perceptible to the majority of end users, it is discernible when the finished tyre is tested by experts.

It is not permissible to supply tyres to one OE manufacturer's homologated specification in place of tyres to another OE manufacturer. To one of SP Tyres' more important OE customers three variants of one tyre size, in the same pattern, are supplied This is for the customer's differing markets so as to help them respond to the varying needs of those markets.

5.4   Compound mixing

The primary rubber processing area (undertaken with internal mixers) making discrete batches has also been affected by the needs for increased flexibility, although to a lesser extent than extrusion. Whilst a proliferation of tyre sizes leads to an increase in components, in the majority of cases these will be made from a common range of compounds. Batch sizes in the order of 200 kg do not normally pose flexibility problems, but to reduce work in progress by reducing production runs does heighten the need for the equipment to rapidly achieve stable temperature settings. Consistency of raw materials in the primary processing of these materials is vital for the successful use in the extrusion and calendering sections of the compounds produced.

6   RAW MATERIAL AND SEMI-FINISHED GOODS SUPPLY

Textile cord preparation has involved the use of large and expensive capital equipment, with a long life cycle. Consequently, this area has not been at the forefront of enhanced flexibility. The prepared material has a limited shelf life prior to rubber calendering. SP Tyres need for small quantities of specialized fabrics, for motorsport in particular, has given their suppliers quite a challenge. Radical solutions are being pursued to overcome the inflexibility of such plant using smaller and vitally different processes. It continues to be vital that the need for flexibility, as well as improved quality, is communicated thoroughly to all suppliers. SP Tyres would not encourage anybody to improve their productivity by installing plant that increases, maybe by days, the size change time that they need in comparison with their existing plant. Instead SP would encourage a search for a solution to both the need to reduce cost and improve response time.

A preference for suppliers within close proximity of their plants, cannot be extended to all materials. Natural rubber is the obvious example. Rubber trees grow in warmer climates than the main tyre producing countries and this situation is not going to change, in the short term at least. Fortunately, whilst the quantities of specific compounds does change, with every change in production size mix, the alteration to the requirements for natural rubber shows less fluctuation, as gains and losses offset each other. In addition, shortfalls can be supplemented from commodity traders, although this is avoided by those companies seeking to maintain ongoing long term relationships with the producers.

Relationships with suppliers must therefore be taken on a case by case situation, with some suppliers able to respond in a fully flexible way, delivering their products reliably with minimum stockholding. With others, where the product quantity involved may be small, (indeed where the tyre industry as a whole may play a minor part of the supplier's turnover), longer lead times may remain the norm, with buffer stocks the only solution to permit quick response to changing demand from the tyre industry's customers.

Whilst some suppliers can and will be relatively close, if it is necessary that in order to meet a specific OE customers needs the suppliers to ship an ingredient such as a single sourced polymer from a distant part of the globe, that must also be done – despite the length of the supply chain. An important aspect of order winning is becoming the ability to respond rapidly to the changing needs expressed by the industry's customers and organization must be adapted to meet these needs. SP Tyres are used to team working across functional boundaries, with a Project Management framework added to make success more certain. At all levels, participants now expect to know not just what is to be done, but also why. Having this knowledge reinforces the identification with the customer and his needs at all levels throughout the company. Ownership of meeting these needs is no longer confined to the salesman, but is shared throughout the process.

7   CONCLUSIONS

Quick response can and is being met, with varying success, by the tyre industry. The companies that will succeed are those that have not just the right equipment, but also practice a philosophy that embraces the need for flexibility with never ending improvement through the participation of all employees to satisfy the ever evolving needs of their customers.

Good communication throughout the process, and dialogue both upstream and downstream so that needs and capabilities are understood and a best fit obtained are vital for success for us all.

Order winning and customer retention are not achieved by producing the right quality and price, as these are now taken as read. The world has moved on, the tyre industry must now respond ever more rapidly both in developing new products and in supplying the varying quantities needed. The success in achieving these goals depends on the people in the industry and their ability to form meaningful partnerships with both our suppliers and our customers, for mutual benefit.

REFERENCES

Bessant, J. R (1991) *Managing Advanced Manufacturing Technology: the challenge of the fifth wave*, NCC Blackwell: Oxford.

Clark, K. B. and Fujimoto, T. (1987) Overlapping Problem Solving in Product Development, Harvard Business School Working Paper, No. 87-048, Cambridge, MA.

Clark, K. B. and Fujimoto, T. (1988a) 'The European Model of Product Development: challenge and opportunity', *Proceedings of MIT International Motor Vehicle Program 2nd Annual Policy Forum*, MIT: Cambridge, MA.

Clark, K. B. and Fujimoto, T. (1988b) Lead Time in Automobile Product Development: Explaining the Japanese Advantage, Harvard Business School Working Paper, No. 89-033, Cambridge, MA.

Clark, K. B. and Fujimoto, T. (1989) Product development and competitiveness, Paper presented at the OECD International Seminar on Science Technology and Economic Growth, Paris, June.

Halberstam, D. (1986) *The Reckoning*, Morrow: New York.

Imai, M. (1986) *Kaizen: The Key to Japan's Competitive Success*, New York: McGraw Hill.

Lamming, R. (1993) *Beyond Partnership: Strategies for Innovation and Lean Supply*, Prentice Hall International (UK) Ltd., London.

Montgomery, D. (1979) *Worker's Control in America: Studies in the History of Work, Technology & Labor Struggles*. New York: Cambridge University Press.

Shigeo Shingo (1985) *A Revolution in Manufacturing – The SMED System*.

Taylor, F. W. (1911) *Principles of Scientific Management*, New York: Harper's.

Womack, J. P., Jones, D. T. and Roos, D. (1986) *The Machine that Changed the World*, Rawson Associates: New York.

13 Changing Methods in the Toiletries Industry: A Case Study

B. Tunmore
Bristol-Myers Company Ltd.

1   INTRODUCTION

Over the last decade there have been many changes within the FMCG arena. Bristol-Myers Consumer Products is part of the American multi-national Bristol-Myers Squibb Corporation ranked within the top pharmaceutical companies in the world. With a turnover in excess of $13 billion and an impressive earnings ratio, Bristol-Myers Squibb has changed dramatically to adapt to these demands.

Within Bristol-Myers Squibb, the UK manufacturing plant located at Cramlington, Northumberland, has embarked upon a variety of innovative schemes to reduce costs and shorten the logistics chain. The plant has 16 high speed toiletries packaging lines, with 200 employees and produces products for the European and Middle Eastern markets, approximately 60 customers. Certified to ISO9002 and Investors in People, and presently working towards TPM and BS7750, the factory is the European manufacturing centre for Bristol-Myers with an award winning level of productivity.

Whilst not constituting guidelines which all can follow, the following article is intended to stimulate thought and give indications of where complications may occur in the path of supply chain reduction. In terms of supply chain reduction and overall logistical improvements, there are various factors and fields which need to be taken into account and, with this in mind, the following will give a brief insight into some of the major areas:

- Quality
- Partnerships
- Financial Implications
- Production.

2   QUALITY

Within a company such as Bristol-Myers, quality is of paramount importance. It cannot be stressed too highly. In other presentations, people may indicate how over a period of time the quality emphasis has changed from that of control to assurance. The important element here is that the more quality assurance the company implements, the less quality control is required. This is done by a variety of ways and means, including operators being responsible for their own quality and stressing that quality is a company-wide need and not something which can be inspected into a product.

Many methods of meeting quality demands have been developed which have their own merits and plus points, such as TQM, Quality Circles and also the philosophy of 'Just In Time'. One of the things Bristol-Myers has found most time consuming is development of quality systems and quality assurance with its own suppliers. Only by embarking upon a constant series of supplier visits, supplier audits, supplier assessments and supplier certification can quality assurance begin to be confirmed. This can be a long drawn out process and it is especially so if the particular industry has not progressed greatly along these paths. To be in the vanguard of supplier certification is a very difficult task indeed. For those companies in industries where the principles are well founded and established, the task is a lot easier. We have to thank our Japanese colleagues for some of this pathbreaking development work.

Once quality from the supplier is assured, then the amount of safety stock held in reserve 'just in case' can be reduced. Companies can now progress along inventory reduction schemes knowing that whatever stock arrives from the supplier is usable. Bristol-Myers itself has changed from a company which inspected, upon receipt, all incoming materials and components to one which has placed the onus of quality very firmly at the feet of the supplier. Providing that the supplier has quality systems in place, for example, ISO9000, ISO9002 or quality systems which have been audited by the customer concerned, there seems little need to replicate the process of quality control upon receipt in the warehouse.

Thus the whole status of quarantine can be removed from all but the most critical or biocritical items.

One word of caution, however, with regard to this method of working is that if a problem is found with a component or raw material, the consequences are far more serious than in the past – namely the non-conforming component will be found on line or during the process of manufacture. Thus the element of increased cost and potential increased consequential loss will be far greater. The problem will also be more visible. Whereas in the past, perhaps, people were used to having a reject area which contained a number of components, they can now be faced with the sight of a production line standing idle because a vital component has been found to be out of specification on line. As the rejected item is also found at the production stage, the time available for reaction is much reduced. Allied to further reductions later on in the chain, the impact can very quickly be felt at the end-customer.

Because of the requirements to have quality products delivered consistently, supplier selection is obviously critical. Great benefit can be had from explaining to present or potential suppliers exactly what is to occur in the future. In some industries this may be a completely new process, in others no change. Bristol-Myers has spent a lot of time presenting and talking to suppliers so that both parties are aware of each other's needs. The benefits to inventory days on hand are worthwhile and much improved inventory turns can result, with beneficial effects on working capital. Component and raw material inventory turns are now into double figures at Cramlington.

3 PARTNERSHIPS

As mentioned previously, the selection process to ensure a quality supplier is critical. Much has been talked about partnership sourcing or 'getting into bed with suppliers'. This, indeed, has many advantages. It should not, however, be assumed that it is suitable for all industries or companies. The respective size of companies is very very important and if the supplier is far larger than the procuring company, this can cause difficulties if the methodologies of working and working practices demanded by the small procurer are at odds with the general terms of the supplier. It is easier to enter into partnership sourcing with those suppliers who are of compatible size to the customer. Bristol-Myers has identified those suppliers with whom it wishes to develop partnerships and has actively resourced materials and components from vendors into a more select core.

For those companies requiring unique moulds or chemicals, this can be a potentially expensive operation. In the case of bottles, for example, if the moulds are presently with one supplier who is not deemed to be the ideal partner, to enable the chosen *alternative* supplier to manufacture may involve capital cost of many thousands to reproduce those moulds. This can make an attractive philosophy a very expensive short-term cost. If, however, the decision is made to select those partner companies, then any new business can be directed to them with no increased cost and, indeed, can result in an overall cost reduction. Where unique moulds or the tools are not required then, obviously, this process is far more simple. Complexities certainly arise when, for example, delivery systems are patented to one supplier or perfumes have been created by a particular company. In these instances, it is not easy to quickly reduce the supplier base, which is an essential factor.

As resource within most companies is limited and far more stretched than in the past, it is not feasible to attempt supplier partnerships with all vendors. Using Pareto analysis, the maximum gain can be achieved from a limited number of suppliers. As the attachment shows, Bristol-Myers used this to good effect, reducing the number of suppliers by over a third in just over two years.

At the same time, it is also prudent to try and reduce the number of items actually purchased from the suppliers used. This can be done in a variety of ways, but obviously globalization of packaging and the subsequent ability to standardize componentry is an important facet in this exercise. As this second chart indicates, Bristol-Myers was able to reduce the number of items purchased by about a third in a similar time period. It is essential to monitor the number of items purchased to ensure that little by little the amount procured does not increase back to an untenable level. This is something that I would advocate is monitored along with the number of suppliers used.

It should not be thought that single sourcing is a panacea for all ailments. Great care needs to be taken when deciding to single source to ensure some form of back-up in the event of a prolonged failure to supply. In certain instances, it may be corporate policy to avoid single sourcing. It can also be disconcerting to find that the chosen partner is proven to be more expensive than a competitor. As partnerships are based on longer term relationships, it does not make good practice to start jumping around between suppliers as the learning curve can be quite lengthy. Nevertheless, there needs to be some form of documentary proof to substantiate why components are being sourced from a more expensive vendor. It is also worth noting that prospective partners never present how they are going to fail in the future! This puts them at some advantage to the existing supplier who has had every failure or weakness observed over the past years.

4   FINANCIAL IMPLICATIONS

Along with the quality supplier who is now tuned into the same wavelength as your company producing a limited or a standardized range of components, there exists the need to exercise financial control. There have been, and there are, companies, in particular those of Japanese heritage, who have made far greater advances in the field of administration than ourselves. The Japanese systems, e.g. Kanban, lend themselves very readily to a more open system of receiving goods and the subsequent payment of invoices. Monthly invoicing may become the norm and with partnerships having a more open approach to accounting or to each other's accounting needs, much more simplified transactions can take place. I must state on this point, however, that this is not acceptable to all companies and those which have financially rigorous audit departments and stringent internal control standards may find their manufacturing departments at odds with the corporate auditors. Great care needs to be taken to ensure that at all times the correct amount of inventory can be accounted for. The representation of inventory on the books of a company is an emotive subject, and having a partner's stock in house without a financial transaction taking place can be hard to justify to those from an audit background. This is certainly one area in which Bristol-Myers has not exploited the potential of partnerships to the full, retaining a need to receive by computer every delivery and de-carding from stock every usage as it happens rather than making a month end adjustment. This is an area which presents untold potential in the future. Whilst working on a product life cycle project, it was noted that certain suppliers did not appear to deliver the amounts stated – for a variety of reasons. Once this is clarified and measures taken to rectify the concern, then Bristol-Myers may take steps to implement this system.

I mentioned earlier the subject of open accounting. In the past, another company's accounts were their business and the open book approach was unheard of. Now, respecting the right to make a profit, it is more acceptable within industry for suppliers and purchasers to openly go through the costing elements which make up the finished price. In this area, one of the most important tools is the

Supplier Agreement whereby vendors and purchasers can agree a methodology of working which is documented, with all assumptions clearly understood by both parties.

In order to make these Supplier Agreements worthwhile, it is necessary to include within the costing formula the various elements which together combined to make up the finished price: how much of the price is based on raw material costs; how much on labour; how much on overheads and how much profit margin is required by the supplier. By analyzing the cost breakdown, agreements can be reached which limit increases on those fixed elements but allow for variability of those elements over which control by the supplier is limited, e.g. raw material costs. This can certainly help during price negotiations where the element to be discussed can be very specific as opposed to a generalistic rise (or decrease).

As mentioned previously, there is also the potential for an alternative supplier to offer much reduced prices. At this point, long-term relationships become more difficult to justify. However, decisions upon supply should be made on best overall value and not just cheapest price.

5   PRODUCTION

Now that the limited number of suppliers are producing quality products at agreed prices, the next area to be investigated is almost certainly the production floor itself. How responsive is the production floor? How customer friendly are the production lines? Is the amount of production geared to what the customer wants or to what the factory wants to produce? In the past, perhaps, the focus was on making products in the most efficient way with regard to labour costs. Labour costs, however, may represent a small portion of the overall cost of the product. That should not be taken in isolation but should also include the unseen cost of large production runs such as huge warehouses, satellite warehouses and, indeed, obsolescence. Large production runs can mean much inventory in a variety of warehouses or countries. Hence the ability to change packaging can become a very lengthy, complex and potentially expensive operation. Although actual production costs may be increased by smaller, less efficient production runs, the *overall* costs to the company may be reduced. These calculations are extremely complex and really do depend on the nature of the business and the ethos of the companies concerned.

Within the European marketplace, Bristol-Myers has moved to dealing with distributors rather than with own company warehousing operations. As stock movements now become financial transactions as opposed to recording inter-company movements, the focus upon inventory holdings becomes that much more acute. There are, however, a variety of caveats which need to be borne in mind when looking at making a production floor more 'customer friendly'.

Perhaps moving away from large production runs to more order-specific volumes is such a new opportunity that the necessary controls are not in place. Certainly Bristol-Myers has found that one of the areas which was overlooked was the necessity and the ability to reconcile production jobs quickly and accurately without interrupting the course of production. To give an example of this, the following figures may be useful. The ideal quantity of a specific pack to be supplied to an individual distributor could represent as little as three minutes running on a production line. Yet accounting and reconciliation, from a batch documentation point of view, would take as long as an order running for a complete day. It is all very well reducing the order size, but care must be taken to ensure that production operatives do not become so disenchanted with the system that they look for ways to circumvent the rules which have been set up.

In certain instances, re-work can be more expensive and time consuming than producing a job from scratch. In order to achieve exact customer order quantities, therefore, it may be easier to produce more and dispose of the unwanted balance. As companies need, and should, devote more time and effort towards protecting the environment, this is not a particularly environmentally friendly method of operation. Job size reduction and the administrative systems which support such reductions must go hand in hand.

Whilst implementing an MRPII style package (BPCS) great care was taken to ensure that all production staff were trained in advance and understood the new systems of working. It was pointed out that the small production runs were not the result of the new computer system, but a response to the real demands of the customer and marketplace. The computer system merely enabled us to service the customer in the way required.

At the same time, we spent a lot of time on changeover reduction projects so that the line down time was minimized. Unlike some industries, e.g. printing, where set up and changeover costs are high and a critical factor, changeover time had been viewed in the past as inexpensive and, hence, relatively unimportant. If a production line for a product range is only operational one week in four, then having a one day changeover is not viewed as critical, as it only concerns one engineer. Through a variety of projects, including multi-skilling, however, changeover times have been reduced from two days to two hours in some instances. This is particularly beneficial on those production lines which have become more heavily loaded due to standardization of componentry and global packaging.

Other methods of reducing cycle times have also been brought into play. Pigging lines and dosing colours/perfumes have enabled between-batch cleaning to be a very quick process indeed. The environmental gains from waste reduction were also an important factor and should not be discounted. This is an area which will increase in importance in the future. Effluent costs and the evermore stringent controls on the disposal of waste make this a priority.

6   SUMMARY

Overall, Bristol-Myers has come a long way in reducing the response times and providing a more efficient customer service. At the same time, the business has changed and the needs of customers have themselves become more demanding.

With the exception of tool or mould building, which is measured in months, the response of the plant is now measured in days rather than weeks. This could not have been done without the involvement and commitment of all concerned. Resource from all quarters has been used. Suggestion schemes with incentives have helped to solicit ideas from the workforce. Project teams under a variety of headings, e.g. Working Capital reduction, have each contributed towards the overall goal. Suppliers have been involved and, indeed, there have been exchanges of engineering knowledge. The reductions would not have been possible without a responsive computer system. Whilst implementing the system, we were able to re-evaluate all of our practices and challenge the need to carry out each and every one and, if proven to be so, did it need to be in that format, e.g. by doing this we were able to develop a system which eliminated a previous 'black hole', that of stock in Work in Progress.

Taken in isolation, each project or achievement would not have made the plant more responsive but, when implemented collectively, the benefits are real and there for all to see.

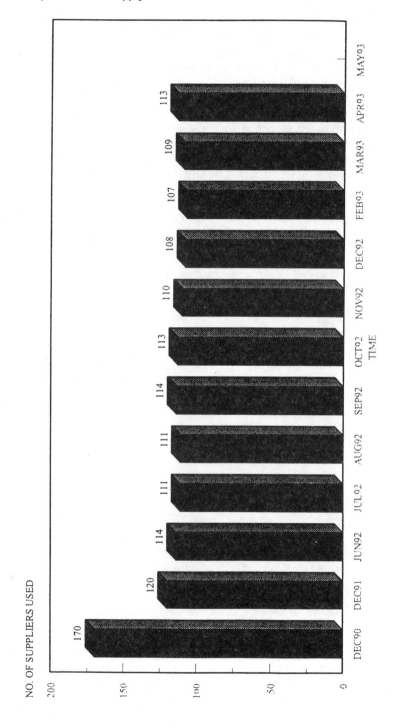

FIGURE 13.1. Performance improvement: number of suppliers used.

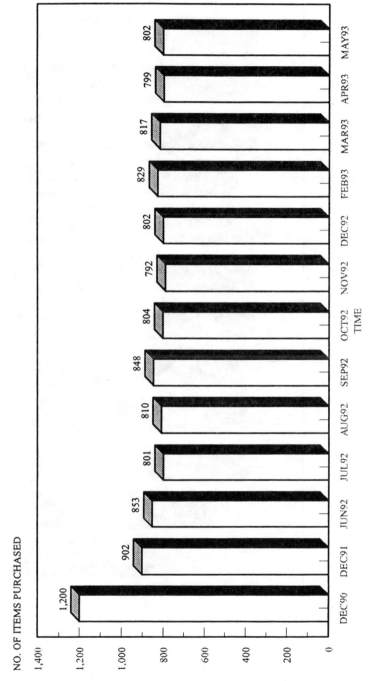

FIGURE 13.2. Performance improvement: number of items purchased.

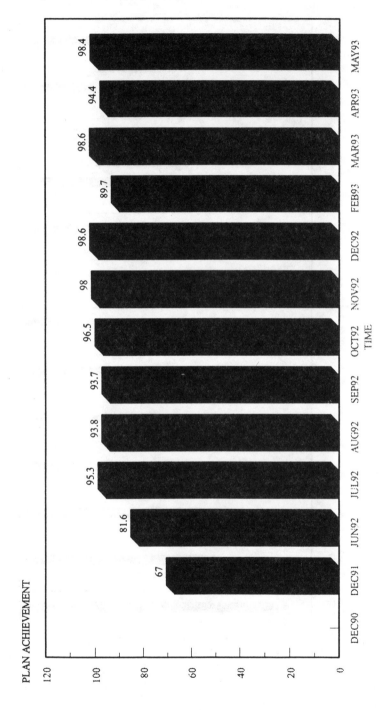

FIGURE 13.3. Operational performance improvement: plane achievement.

14 Time Based Distribution

M. Abrahamsson
University of Linköping and Logicent AB
Sweden

1 INTRODUCTION

Physical distribution of industrial goods is an area that has not changed during the last few decades. Production oriented companies usually have a long distribution channel to break down large production batches, step by step, warehouse by warehouse, to a product-mix demanded by the customers. A sales-oriented company uses a wide distribution channel, with many sales offices and warehouses, to be geographically close to the customers.

In such decentralized and traditional distribution structures, each link in the distribution chain usually manages both sales and warehousing. The production unit delivers to a national warehouse, which delivers to local sales offices that manage the deliveries to the customers. In consequence the distribution system includes a lot of warehouses (Figure 14.1).

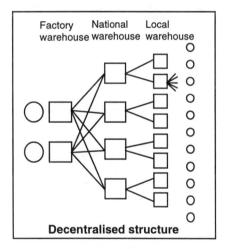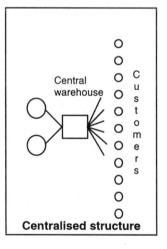

FIGURE 14.1. The change from a decentralized to a centralized distribution structure.

This article follows a simple formula for strategic change WHAT × HOW = RESULTS, which indicates that it is not enough to know what to do if you do not know how to do it. The article presents WHAT to do to change the distribution system to a system with distribution from one or two central warehouses in Europe direct to all individual customers, 'Time Based Distribution' (Abrahamsson, 1992), HOW to do it and the RESULTS of such a change.

The article is based on a study of three international Swedish companies for the time period of 1980 to 1992 – Atlas Copco Tools AB, AB Sandvik Coromant and ABB Motors AB – and their successful redesign of the distribution to the European market. The customers in Europe now get their deliveries direct from one (in the Sandvik case two) central warehouse. The companies are all manufacturers of industrial goods and they all have sales-companies in almost all the European countries.

Atlas Copco Tools manufacture hand held pneumatic tools. The assortment contains about 1000 different products. A customer's order normally contains three items, and the company handles about 1000 customer orders daily. About 80% of the annual sales were to European customers.

Sandvik Coromant manufactures cutting tools for industrial use and the assortment contains about 18000 items. The average customer order contains 2–3 different products and about 3–4000 orders are handled daily. Of the total sales 97% was export, mainly to Europe and the US.

ABB Motors produce standardised and customised electrical motors, from 0, 18 to 400 kW. The assortment contains about 1000 different standard motors, and a customer order contains two different types of motors. The number of customers in Europe was about 600, which was equal to 50000 customer orders annually.

2   WHAT TO DO

According to the marketing channels theory the design of a distribution structure is based on the geographical distance between the producer and its customers, according to the customers demand for products and services.

Because of the gaps in the channel (time gap, geographical gap, quantity gap and variety gap), the producer and the customers operate almost independently. The producer's large production batches do not match the customer demand for small quantities and specific assortment. Similarly the time and place of production does not match the time and place of the customer demand. Therefore the middlemen are expected to create utilities in place, time, quantity, assortment and possession (Kotler, 1988; Stern and El Ansary, 1988).

According to the marketing channels theories the distribution channel will contain many warehouses if (Jackson *et al.*, 1982):

- The number of customers is large.
- The customer-structure is geographically spread out.
- The industrial concentration of the market is small.

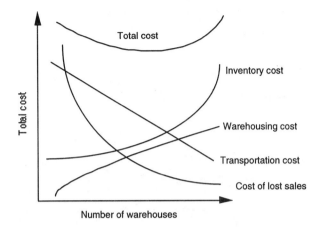

FIGURE 14.2. Total distribution cost. Source Coyle, Bardi and Langley (1988).

- The customers purchase small volumes.
- There are many competent middlemen available.
- The products are standardized.

The logistics theories differ from the marketing channel theories in two ways.

In the logistics theories co-ordinating functions are important and considered as necessary to reduce costs by trade-offs. Increased costs in one function can result in decreased costs in other functions – thereby reducing total distribution costs (Christopher, 1986). The number of warehouses is a result of a total cost analysis (Figure 14.2). Where the total physical distribution cost is a function of inventory cost, warehousing cost, transportation cost and cost of lost sales (Coyle *et al.*, 1988).

As in the marketing channels theories the geographical distance to the customer is important. The number of warehouses is considered to effect the customer service. That's why the cost of lost sales is expected to increase when the number of warehouses is reduced. Another factor that effects the number of warehouses in the logistic theories are lead-times (the time from received order to a complete delivery to the customer). If the demand for short lead-times is high, or if the customer's orders are hard to predict, then another warehouse in a closer location to the customers is recommended (Christopher, 1986).

2.1 The driving forces was rationalization or market orientation

In the beginning of the 1980s Atlas Copco Tools had seven manufacturing plants in Europe, each with a warehouse for finished goods. They had two central warehouses, one in Sweden and one in Finland. In each country there was a sales company with a warehouse. Some countries were divided into regions, in Germany for example there were five regions with local warehouses. The local warehouses

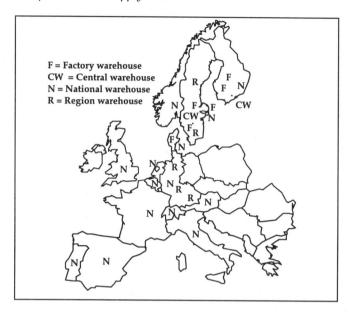

FIGURE 14.3. Atlas Copco Tools distribution structure before the central-ization.

were refilled from the national warehouse, which was refilled from one of the central warehouses, which was refilled from the factories stock of finished goods. The distribution structure is shown in Figure 14.3.

Sandvik Coromant and ABB Motors had a distribution structure similar to Atlas Copco Tools in Figure 14.3. Sandvik Coromant had two central warehouses in their distribution system, one in Sweden (close to the main production unit) and one in Holland. The assortment of stock in the sales companies contained about 4 – 6000 items. The distribution strategy was to be geographically close to the customers.

The assortment in the ABB Motors local warehouses was limited and each production unit only manufactured a limited part of the assortment. Consequently each sales company frequently ordered motors for refilling the warehouse from all six production units in Europe. Because of the administration and materials handling in the local warehouses only 30% of the personnel resources at the sales companies could be used for sales activities.

The DDD-system (Daily Direct Distribution) started in 1987. It was the second step in a rationalization programme which started in production in the early 1980s because of poor profitability. The overall motive was to reduce costs. With DDD Atlas Copco Tools delivered all their products from one central warehouse, located in Sweden, direct to the customers in Europe.

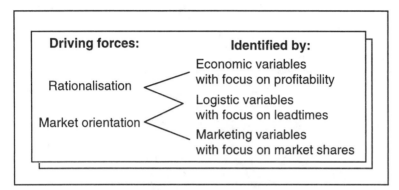

FIGURE 14.4. Driving forces for a change to 'Time based Distribution'.

Sandvik Coromant's change of the distribution system started in 1982 with their DD-system, Direct Distribution of customer addressed deliveries from the central warehouses to the local sales companies. The motive was to increase cost efficiency and profitability, the goal was to reduce tied-up capital with no changes in customer service and market shares. In 1984 Sandvik Coromant began to deliver direct to customers from the two central warehouses, and the DDC-system (Direct Distribution to Customer) was born. The driving force for both Atlas Copco Tools and Sandvik Coromant were rationalization with focus on profitilability (see Figure 14.4).

The triggering factor for ABC Motors' centralization of the physical distribution was a market research amongst 600 customers in Europe. The results showed too high costs for sales and distribution, poor delivery performance and that the company was too production oriented with limited knowledge of the customer demand. But most important, it also showed that the customers were not satisfied with any of the suppliers of electrical motors.

The driving force for ABB Motors' 'Prime Move System' was market orientation in order to increase market shares by improving the delivery performance and to set free resources in the sales companies for marketing activities (see Figure 14.4).

The new distribution structure was similar for all three companies, with only one or two warehouses for the European market. In the beginning Atlas Copco Tools used their existing warehouse in Sweden. However, in January 1992 they moved the warehouse to Belgium (see Figure 14.5).

Sandvik Coromant still have two warehouses in their system, one in Sweden and one in Holland. In the DDC-system a customer order normally contains products from both the Swedish and the Dutch warehouses. The shipments from Sweden go by air to a break point in Frankfurt, while the shipments from Holland arrive by truck. The local transport is different in each country, mostly it is by mail or by truck. The new distribution structure is similar to Tools structure in Figure 14.5.

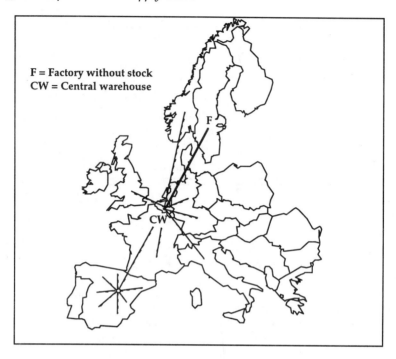

F = Factory without stock
CW = Central warehouse

FIGURE 14.5. Atlas Copco Tools distribution structure after the centralization with only one warehouse for the European market.

ABB Motors located their central warehouse in Germany because Germany was the most important market. To convert fixed costs to variable costs, ABB Motors decided to buy the physical distribution services. In 1988 the German forwarder Schenker established a central warehouse in the area of Düsseldorf dedicated for ABB Motors.

3   HOW TO DO IT

In James Cooper's article 'The Paradox of Logistics in Europe' (Cooper, 1991), he argues that there are several major cost implications of concentrating inventory in a few warehouses, e.g. less tied up capital, decreased costs of operating warehouses and administration of the physical distribution. He also notices that many companies, wishing to develop a 'pan-European logistics system' with fewer and larger storage points, are not making any significant progress to reach this goal.

This implies that the potential effects of such a redesign of the physical distribution is well known but the experience on how to do it is low. The HOW-question can be divided in three phases:

- **Design and planning:** which is the necessary pre-work and answer the question how to design a Time Based Distribution system.
- **Organization:** which focus on how to organize the responsibilities.
- **Implementation:** which answer the question how to get the distribution system running.

3.1 Design and planning

In the design and planning phase there are two important tools. First the *lead time*, defined as the total time from that the customer place his order to he receive the goods (Stalk, 1988). The lead time is the tool to determine the distribution structure, or the number of warehouses. The second tool is the *information system*, which is the tool to decrease the lead time.

Before the change the studied companies measured the gap to the customer geographically in terms of distance (miles, or kilometres). With the new distribution strategy they focus the gap in terms of lead time. In Time-Based Distribution it is more important to deliver the goods to the customers within a specified time, e.g. 48 hours, than it is to have warehouses geographically close to the customers.

In the traditional models the distribution structure is a function of all variables represented in the marketing channels and logistics theories. In Time Based Distribution the lead time is the only variable to consider to calculate the number of warehouses. The first thing to do in the design phase is to identify the lead time demanded by the customers. If it is possible to reach the customers within the requested lead time, e.g. 48 hours, with only one warehouse and with regular transport systems than one warehouse is the best structure in terms of costs and customer service. If you cannot reach them within 48 hours you have to add another warehouse to the structure and so on.

Time based distribution is very much based on modern information systems. The information system is the tool to decrease the lead time to a minimum. The lead time can be divided in the physical lead time and the administrative lead time. With the latest information systems the administrative lead time, which in many cases is the longest, can be reduced to almost zero, e.g.

- Order handling
- Order processing
- Picking lists
- Freight documents
- Customs documents
- Delivery reports
- Invoicing.

The lead time from Sandvik Coromant's local sales companies to customers was normally 10 to 30 days. More than 90% was administrative time. With modern information technology the administrative lead time was reduced to less than one day, and the total lead time to 24 hours.

3.2   Organization

A necessary organisational approach is to get an 'helicopter-view' on the organization and on the logistic activities. Time based distribution is not functional logistics or regional – or national logistics but international and global logistics. If you do not have such a helicopter perspective you cannot see the potential benefits because you get stuck in problems on departmental or local levels in the organization. Problems that will show up as a result of barriers in the organization against structural changes.

But with an helicopter perspective you will see that the global logistics in most companies, representing different industries, are very similar up to 70–80% of the logistics activities. You will also see that the potential benefits are in the change of the distribution structure, not in rationalization or computerization of existing logistic functions.

Time based distribution affects many different departments in the company in different ways. To get the project running it has been shown successful to set up a small project organization, with members from the departments of logistics, marketing and sales, production and computer systems. If you have members from those departments in the project organization from the beginning the changes to get project accepted in the most critical departments will increase. It is also important to get full support from the top management from the beginning. Because support from the board of directors is necessary for the project to be accepted within the organization.

Another organizational issue is the responsibilities. In Time Based Distribution the physical flow will be separated from the sales activities. The physical distribution will be centralized to get full control over the logistic activities and the sales activities will maintain local. This implies a change of the responsibilities in the logistic chain.

A successful model has been to create a 'logistics control function' with responsibilities for customer service and logistic control (inbound logistics from the production units to the central warehouse, materials handling in the central warehouse and outbound logistics to the customers). However, the production unit has the responsibilities for tied up capital for items in stock at the central warehouse. Because the second best way to reduce tied up capital and decrease the stock levels (the best way is the decrease the number of warehouses) is to force the production to be more flexible and refill the warehouse more often.

3.3   Implementation

In order to disarm the largest enemies to a redesign of the distribution system, which very often is your own sales companies, there are two important guidelines for the implementation phase. First to use fast transports to initially secure the customer service. Initially most of the shipments from Atlas Copco Tools central warehouse in Sweden to customers in Europe were by air freight. Both to be sure of the delivery performance and to convince the local sales companies that they did not need stocks of their own. Now the system is up and running, 95% of the shipments are by truck. With all volumes for the European market concentrated to one single warehouse full truck loads can be used from the warehouse to a local break point in each country. From the local break point the goods are delivered to the customers through a forwarders domestic network or by the postal service.

Second, use the first market as a show case. To get a Time Based Distribution system running it is very important to make sure all systems, e.g. information system, materials handling system and transport systems are working properly. It is also important to be able to show the effects in economic terms from the first market before the rest of the markets is implemented.

4   THE RESULT IS INCREASED COMPETITIVE ADVANTAGES

Atlas Copco Tools former distribution strategy was influenced by the traditional theories. It was designed to be geographically close to the customers. The aim was to have a complete assortment in the local warehouses. However, the delivery performance (number of items ordered available in stock) was only 70%. If the items could be delivered from stock the lead time was two days. However, the average lead time over all orders was about two weeks. In 1986 inventory was 29% of the sales, and the value of the sales companies stock was $ 9 million. The identified effects of the change to Time Based Distribution were:

- Reduced inventory by 1/3 from 29% to 20% of sales.
- Reduced variable distribution costs by $4 million annually.
- Reduced average lead time to customers from 2 weeks to 24–72 hours (depending on the market in Europe).
- Increased delivery performance from 70% to 93%.
- Reduced number of employees in the central warehouse from 40 to 23.

The pattern for the other two companies was the same. For Sandvik Coromant the identified effects of the change to Time Based Distribution were:

- Reduced inventory to one third, or by $80 million.
- Reduced tied-up capital by $16 million annually.
- Reduced lead time from order to local break point from 10–30 days to 24 hours in Europe, with a reliability 99%.
- Increased delivery performance from 90% for the locally stocked items to 90–99% for the whole assortment (90% for tools, 95% for bits and 99% for spare parts).

- Reduced number of warehouse personnel by 55 persons (1986).

Because of ABB MOTORS limited assortment in the local warehouses and the complex distribution system only 30% of the orders were completely delivered from the local warehouses. The average lead time for complete orders was 2–4 weeks, and the total distribution and sales costs were 35% of sales. With Time Based Distribution they:

- Reduced total distribution and sales costs, from 35% to 20% of sales
- Reduced fixed and variable distribution costs
- Reduced lead time from 2–4 weeks to 24–72 hours
- Increased delivery performance (number of items delivered from stock) from 50% to 95% for the whole assortment.

Influenced by Perssons (1990) classification of logistics in its importance as a 'cost driver' and as a 'unique driver', who in turn was influenced by Porters (1980) approaches to competitive advantages through overall cost leadership or differentiation, the effects could be expressed as competitive advantage. In the three cases the logistics has been important both as a cost driver and a unique driver and the logistics has been flow oriented, Figure 14.6.

In terms of competitive advantages the effects of Time Based Distribution could be described as logistics cost advantages for the selling company, and added values for the buyer. The effects identified in the three case studies are summarized in Figure 14.7.

As expected the companies with a driving force to reduce inventories did so. However, the company with a driving force to increase market orientation also achieved a remarkable reduction in terms of rationalization. According to the tradi-

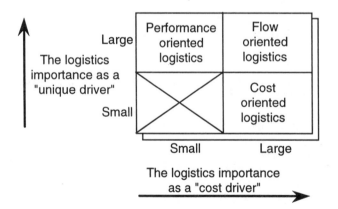

FIGURE 14.6. The logistics importance as 'cost-driver' and 'unique-driver'.
Source: Persson (1990).

tional logistics theories the transportation costs were expected to increase considerably. Yet the transportation costs did not increase in any of the three cases. The reason is a complete assortment in the central warehouse in combination with a smooth flow of deliveries from the warehouse which made stock control much easier. As shown in Figure 14.8 there is a clear connection between centralization of the physical distribution and decreased inventory costs, with constant transportation costs.

Savings in integration of functions were partly a result of modern information technology. A computerised information system was absolutely necessary to implement Time Based Distribution, which makes it possible to integrate e.g. the order handling system with production planning. An order to the central warehouse is also a signal to the production planning system to refill the warehouse. But it is important to use the information system as a catalyst to change the distribution structure. Not to computerize existing distribution structure. Because than you will cement the existing structure, which will be even more difficult to change.

Another important effect of centralization was separation of the sales function from physical distribution. Physical distribution is centralized to achieve economy of scale in materials handling and transportation, and the sales function is still local for best customer service and support. Because then personnel in the sales companies did not have to handle and control their warehouse, they had more time for marketing and sales activities.

Logistics cost leadership	Logistics buyer value
Fixed distribution costs: - Decreased costs for personnel, warehouses and administration	**Lead times:** - Shorter and more reliable lead times for all markets and for all products
Variable distribution costs: - Reduced inventory costs - Constant transportation costs	**Delivery performance:** - Increased on-time deliveries - Complete orders to the customers
Savings in integration/separation: - Sales function separated from the materials flow - Centralised control of the materials flow - economies of scale - Integrated distribution functions	**Differentiation:** - Customised distribution to different groups of customers - Increased flexibility
Savings in learning costs: - Faster introduction of new products in the assortment	**Customer information:** - Faster and more reliable information to the customers about dicrepancy

FIGURE 14.7. General effects of Time Based Direct Distribution.

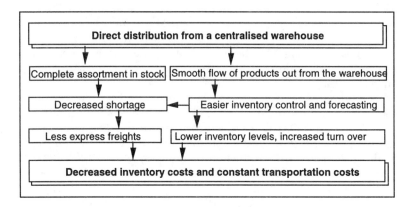

FIGURE 14.8. Connections between centralized distribution and inventory/transportation costs.

The savings in learning costs were obvious in Sandvik Coromant AB. They annually introduce several products. With only two warehouses for the European market it is possible to replace the old products with new ones in a short time and on all markets simultaneously.

On the buyer value side of Figure 14.7 the effects on lead times and delivery performance were most obvious. The lead times were no more than 72 hours in any market and, which is most important, the lead-times were reliable for all the products and for all markets. Again the key was a complete assortment and the smooth flow of deliveries from the warehouse, which makes it easier to control the availability of products in the warehouse.

Some representatives of the companies in the study had the opinion that Time Based Distribution made it easier to give the customers correct and fast information about discrepancies in the deliveries. Because of the flexibility in the system, direct distribution from a central warehouse also gave them opportunities to differentiate the distribution to different groups of customers.

4.1   It is time to up-date the traditional theories

Opposed to the traditional theories, the companies found their best distribution system with only one or two warehouses in the system. They focused on delivery times instead of the geographical distance to the customers, a factor which is not considered as significant in the traditional theories.

In the three case studies the total distribution costs decreased with a reduction

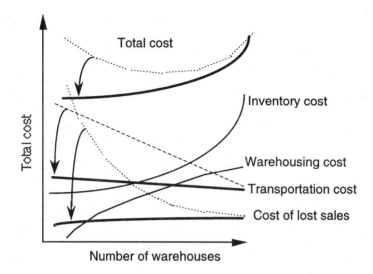

FIGURE 14.9. Changes in the total distribution costs model.

of the number of warehouses in the distribution structure. The inventory costs decreased, the transportation costs were constant and the sales increased rather than decreased. It is the opposite to what was expected in the traditional total distribution costs model in Figure 14.2.

The key to these effects is a complete assortment in stock, which is possible with only one warehouse, and a smoother flow of products out from the warehouse, because of the large number of small deliveries to a large number of customers. The traditional models do not consider the importance of these variables to the design of a distribution system. The variables have an impact on two of the curves in the total distribution costs model – the transportation costs and, because of shorter and more reliable lead-times, the curve of lost sales. It shows that the total distribution costs will increase by the number of warehouses, as shown in Figure 14.9.

4.2 CONCLUSIONS

In the three case studies the distance to the customers in time determined the distribution structure. Atlas Copco Tools and Sandvik Coromant initially used air freight to reach the customers in time. ABB Motors located their warehouse in Germany, to reach the European market in 24–72 hours by truck. The importance of the lead time in the distribution structure, and because the companies studied decreased their distribution costs, indicate that the lowest total distribution costs will be reached with *as few warehouses as possible and requested lead time as a prerequisite.*

The clear effects of the change to 'Time Based Distribution', both in logistics costs leadership and logistics buyer values, show its potential to increase the competitive advantages of the companies. Not only by cost leadership but also in differentiation. The three companies in the study were all pioneers. Recently they have influenced other companies, representing other industries, to change their distribution in the same direction. Time Based Distribution is a good example on the potential to brake traditional patterns and change the distribution structure. Because the potential benefits are in the re-engineering of the business processes, not in rationalization or computerization of existing routines.

REFERENCES

Abrahamsson, M. (1992) *Tidsstyrd direktdistribution – Drivkrafter och logistiska konkurrensfördelar med centrallagring av producentvaror*, Studentlitteratur, Lund.

Christopher, M. (1986) *The Strategy of Distribution Management*, Heineman Professional Publishing, Oxford.

Cooper, J. (1991) The Paradox of Logistics in Europe, *International Journal of Logistics Management*, 2(2), pp. 42-54.

Coyle, J., Bardi, E. and Langley, Jr. J. (1988) *The Management of Business Logistics*, West Publishing, New York.

Jackson, D., Krampf, R. and Konopa, L. (1982) Factors That Influence to Length of Industrial Channels, *Marketing Management*, 11, November, pp. 263- 268.

Kotler, P. (1988) *Marketing Management*, Prentice Hall, New Jersey.

Persson, G. (1990) Logistik som konkurrensstrategi, BI, Working Paper 1990/2, Oslo.

Porter, M. (1980) *Competitive Strategy*, The Free Press, New York.

Stalk, G. (1988) Time – The Next Source of Competitive Advantage, *Harvard Business Review*, 66(4), pp. 41-51.

Stern, L. and El-Ansary, A. (1988) *Marketing Channels*, Prentice Hall, New Jersey.

15 Integrators: The Challenge of Changing Logistics Structures

T. Keating
UPS Europe
Former Director of Logistics

I would like to explain how an integrated operator is approaching the task of establishing a network of services throughout Europe and how this fits into the various logistic changes that are currently taking place.

There is one thing on which we all agree – and that is that there are major changes taking place in logistics activities worldwide as companies try to draw the correct balance of where they should source raw materials, where they should manufacture or assemble their products, whether goods should be kept in the central, regional or local distribution centre (if indeed any distribution centre) and how this best fits the need to respond to customer requirements in an increasingly competitive environment.

Correctly analyzing the supply chain is a complex task, if all the advantages and savings are to be achieved. Indeed I believe most companies, if they are honest, would admit that it often turns out to be a great deal more difficult than they first thought. Logistics activities impact most departments inside a company often there are interesting conflicts between internal profit centres as to what action should be taken.

However big the challenge, there is no doubt in my mind that trying to achieve the correct logistics structure provides one of the great corporate opportunities of the nineties, with important financial savings to those who get it right.

What exactly is driving this search for new logistics structures? Is it because we now operate in a global marketplace, or perhaps due to the increased competitive environment in which we find ourselves? Is it due to the regional, political and market changes such as European borders coming down or NAFTA (North American Free Trade Association) being created? Is it due to the latest research in logistics theory and thinking, or indeed the development of new and more reliable door-to-door delivery options?

I believe that all these factors are contributing to the dramatic changes that are currently taking place and the most important point is that all of them are happening at the same time, which is giving greater impetus to the search for the optimum solution to this fascinating logistics question.

I have personally been involved in international transportation now for 36 years and think honestly say that this is probably the biggest period of change and opportunity I can personally remember.

The whole question of logistics is, of course, far more complex than just transportation or distribution. On the other hand, I would argue that transport is one of the most critical elements to developing a successful logistics solution. Depending on the value and structure of your operations, transport can be up to 40% of the total logistics costs. But without that critical on-time delivery and control from the transport point of view, many of the other important financial savings cannot be achieved. Without a totally reliable and cost-effective delivery system that is monitored and controlled throughout, you simply cannot implement an effective JIT manufacturing system and close down that local warehouse or reduce stock in a regional distribution centre to the levels needed to obtain those important savings in inventory levels.

Until relatively recently, the traditional structure of the transport and distribution industry has been reasonably well defined. Most companies knew whether they were a courier or forwarder or NVOCC operator or broker. Some of the larger conglomerates of ten embrace one or more of these individual functions, but if so, they were nearly always kept as distinct and separate divisions within the overall corporate entity.

Established relationships developed between these various sectors, for example between airfreight forwarders, airlines and brokers or between NVOCC operators and shipping lines. Sometimes these were based on rather a 'love-hate' relationship, but usually in the end, they existed because each sector needed the other to carry out the movement to the final destination.

This traditional structure is now beginning to change. It is no longer quite so clear whether you are a forwarder or an airline because forwarders, for example, now operate aircraft. Or indeed whether you are an integrator or an express operator. With the recent changes in the European post office arena it is sometimes difficult to know whether you are a postal service or an integrator.

Of course, in some cases these divisions will remain, for example in the movement of bulk commodities by sea or rail, or where specific commodities require specialist knowledge and so on. In many other areas, however, there is an increasing convergence between these activities and the previous clear divisions are becoming more and more blurred.

Take, for example, a company sourcing raw materials in Canada or Latin America and moving these by bulk shipping to the Far East for manufacture or assembly into electrical products. These products might then have been moved by sea or traditional airfreight services to Rotterdam or Schiphol, where they would have been cleared through customs and eventually been sent by individual FTL or LTL services to individual country warehouses. The final customer would be supplied from these warehouses, possibly by a local postal or domestic parcel operation. The overall lead time from manufacturing in the Far East may well have been counted in months, but many companies are now able to develop structures that will result in a lead time of days rather than weeks by co-ordinating the transport from the Far East and supplying direct from one regional European distribution centre or even direct from a central distribution centre located in the Far East.

To achieve this, most companies are beginning to demand single carrier responsibility for their transport and distribution, at least for each major leg of the shipping process.

In many cases, transportation costs have already been dramatically reduced through the introduction of these integrated door-to-door services, which are usually more reliable with time-controlled delivery and simplified documentation, offering overnight, 24, 48 or 72 hour service options. They are usually backed by computerized tracking and tracing system and with simple all-inclusive rates structures. The service is efficient and cost-effective.

These new networks are now developing both in Europe and worldwide and with continuing moves towards one single market in Europe we can expect these changes to continue. But just how fast will they develop and what impact will they have on logistics and indeed how cost effective will they be?

In order to answer some of these questions, I believe it is interesting to look at some of the developments that have taken place in the USA, where similar changes over the past decade have created a rapid growth in the evolution of integrated operations.

I must stress I fully agree that there are many important differences in the way business is carried out in the USA versus Europe. We are different countries with different cultures, languages, currencies and so on. Nevertheless, there are also many similarities and I still believe much can be learnt from the US experience.

The population of Western Europe is greater than that of USA and, if Eastern Europe is included, the total potential market is also far larger. The GNP of the two areas is very similar and although the distribution of industry is not the same (if anything industry is more fragmented and distances are greater in the USA), Europe should have a distinct advantage.

The USA is already effectively a single market. Although there are legal differences between the states with different sales taxes and so on, basically there is now fairly open competition throughout the USA as far as a transport is concerned. This single large domestic market contrasts strongly with Europe, where transport has tended to develop in a fragmented manner domestically within each country. Only relatively recently have true pan-European networks, both by ground and air, emerged.

To what extent the development of these integrated operations in the USA has driven the trend towards centralized distribution or to what extent the trend towards centralization has developed the integrators' networks can be endlessly debated, but the fact is, they both need each other.

The result has been the development of large nationwide parcel operators such as UPS, Federal Express and Roadway. Outside of the USA many people do not realize the size to which these integrated operations have now developed. I am sure you will understand if I use UPS as an example.

The following statistics will give you some idea of the size and scope of UPS' operations. We handle more than 12 million parcels every day, about a third of which are overnight deliveries. UPS' 1995 turnover exceeded US$ 21 billion and we operate 130,000 vehicles and employ a workforce of over 335,000 people.

As a result of the huge sums being invested in our international operations, I am frequently asked if UPS is still profitable. As we are a private company, owned entirely by our 25,000 managers worldwide, financial results are not normally published, but I can tell you that the pre-tax profit for 1995 was $1 billion, even after absorbing our current investments in our new international services.

The reason for stressing the size of the operation is not to mesmerize you with numbers, but to emphasise the huge economies of scale the parcel industry has achieved in the USA. Individual unit costs reduce dramatically as a result of larger scale operations and yet at the same time on-time delivery performance is much higher than currently achieved in Europe, with levels of 99.7 or 99.8% being achieved on standard services.

Louisville air hub, for example, is based on 250 acres of land with nearly 250,000 square metres of buildings, of which the Sort facility is capable of handling 160,000 packages per hour. Airhubs in Europe, at the present time, are far smaller, but as they increase their throughput there is no reason that comparable economies of scale should not be reached over here too.

Similarly, the type of aircraft used within the USA to achieve overnight parcel deliveries are much larger and more cost-effective to operate than those currently used within Europe. Only 7% of our parcels move by air with the other 93% moving by ground, but in order to move this 7% we own one of the world's largest airlines with a fleet of freighter aircraft which even the larger scheduled airlines would find difficult to match.

It is not just in the air but also on the ground that the size of the handling operations in the USA produce economies far greater than those currently available within Europe Standard pick-up and delivery centres within the USA are much larger than those within Europe. But, again, as similar operations develop in Europe, important economies of scale will follow.

The larger operations also offer high levels of efficiency. If we take, for example, the simple question of pick-up and delivery in the USA, UPS achieves an average of 145 stops a day from each of its delivery vans and in urban centres the average number of stops per day can rise to well over 200. This naturally has a major impact on the cost per stop which once again is far lower than those currently achieved within Europe.

Similar consideration is given to the design and maintenance of vehicles in order to achieve better reliability and a longer vehicle life. A typical UPS vehicle life has a life of about 18 years and this can extend up to 25 years. Each vehicle would cover over 700,000 miles, and in some cases as much as 1,000,000 through efficient maintenance and planning before it is replaced.

Equally important, of course, have been the major improvements in technology, which over the last few years has changed the way that so many parcel systems operate.

At UPS previously hand-written pick-up and delivery records have now been replaced by portable hand-held computers in which the driver keys each activity and can even digitally capture the customer's signature. This is electronically recorded and transmitted direct from the vehicle cab to the UPS mainframe via new cellular data networks that have been established coast-to-coast. This enables POD information together, with all other pick-up and delivery records, to be accessible immediately by both shippers and consignees.

This information, which is keyed at origin, giving the shipper's name and address as well as the consignee's name and address, type of goods, value of goods, etc., is then transmitted worldwide via the two systems – UPS' 'International Shipment Processing System' (ISPS) and the Parcel Tracking System' (PTS)

This one input provides information on the various road and aircraft manifests throughout the movement and as these are co-operative systems, expected arrivals can be pre-alerted to each transit point.

At destination the data is automatically passed to the various Customs authorities in advance to provide full customs entry and statistical information. The final destination centres also know exactly which parcels to expect through this pre-alert system, and they are able to plan their days operation before any of the shipments physically arrive.

Once the parcel has been delivered, by capturing the name of the person that signed for the package together with the date and time, an effective POD becomes immediately available on a worldwide basis.

Another current development, both in the USA and Europe, is electronic mapping which enables pick-up and delivery efficiencies to be dramatically improved. These systems display detailed maps of every street and highway in towns in the USA and Europe and, with pinpoint accuracy, show the location of vehicles and customer addresses, which yet again enables costs to be reduced with more effective routing of vehicles.

EDI links are also now becoming regularly established with key customers and there are a variety of ways of ways in which shippers have software and hardware on site which can prepare documentation, print waybills, produce daily, weekly, monthly analysis of their traffic, as well as trace shipments and show who exactly signed for their package and when.

I have particularly emphasized the economies of scale, efficiency and technology produced by these new integrated operations in the USA because I believe they are the key factors that are the large differences in cost that at present exist between USA and Europe in the movement of parcels.

If you look at tariff levels of integrators in the USA and compare them with those in Europe, there are huge differences. If you look at the rates in the USA, a next day a.m. delivery would cost about $20 for a three kilo package and $39 for a ten kilo package. If that movement went by ground, keeping to the same transit time, the cost could be as low as $2.80 for a three kilo package and just over $5 for a ten kilo package. And I would stress that these are full tariff rates before any discounts have been applied.

In addition, rates in the USA have been continuing to reduce in real terms over recent years. For example, between 1982–1990 UPS rates increased by 15.5% but if inflation is taken into consideration, the underlying trend has been a decrease in rates of 27% in real terms. I have talked a lot about the USA and the main question for us in Europe is to what extent are we really going to benefit from these developments and how quickly will similar networks and costs arrive over here?

As the borders increasingly come down and more companies regard Europe as one market instead of individual country markets, we should start to benefit from the changes that have already taken place within the USA.

Over the last few years vast sums of money have been invested in establishing similar integrated parcel networks throughout Europe which provide a range of document, parcel and freight services to every address in every country. Several integrators now offer a range of time-controlled 24, 48, 72 hour services by either air or ground at different costs depending on the speed required.

To deliver urgent shipments throughout Europe, particularly when next day a.m. deliveries are required, it is necessary to develop an air network based on a central hub. UPS, for example, has developed an air hub based in Cologne, Germany. We fly an extensive fleet of aircraft nightly to all main countries in Europe with flights that link in with the USA air netwrok.

For less time-sensitive shipments with transit times of perhaps 48 or 72 hours (but still time-definite) a lower cost option is to establish a pan-European road network.

UPS has invested heavily in developing a comprehensive ground network in Europe. An important factor in its success is the ability to co-ordinate these networks wherever necessary to provide a range of joint air and ground options.

There will, of course, continue to be a strong demand for local domestic distribution within each country in spite of the Custom changes. A high volume of traffic moving domestically means that the local pick-up and delivery costs can be kept low, and by co-ordinating these domestic pick-up and delivery services with the pan-European networks, overall costs can be even further reduced. Once again, however, this means developing a totally integrated portfolio of both domestic and pan-European services.

Several integrators are developing their own total European networks, and to date UPS has invested over US$ 1 billion setting up its current infrastructure, with a further $1 billion investment planned for the next five years. I believe UPS already has the largest network of any operator in Europe, handling over 600,000 parcels every night in Europe. In addition to our separate pan-European air and ground networks, we now also have major local domestic networks in Germany, Italy, Spain, France, Belgium, Netherlands and UK.

It is the ability to co-ordinate these different services into one total network that is enabling a range of new services to be introduced, but at the same time maintaining the overall control and a co-ordinated tracking and tracing system throughout.

For example, larger shippers moving higher volumes of parcels and freight often prefer the concept of having a lot of smaller individual parcels combined as one large shipment for moving from one country to another and then dropping the individual parcels into the domestic distribution network of the destination country. We call this 'drop shipping'. Costs are saved as there is only one Customs clearance on arrival and even where borders have come down, there is still the advantage of only one payment of VAT and one VAT and statistical document to prepare, but with individual parcels moving directly to their final consignees. The emergence of fully integrated time-controlled delivery will have an important effect on all aspects of material sourcing, manufacturing and assembly locations, warehousing and inventory systems, as well as the location of central or regional distribution centres – all dramatically affecting the overall cost and service to customers.

Another specialist service introduced by UPS in 1992 included the provision of bonded and non-bonded distribution centres which linked into its pan-European network. This service is of interest to European companies, but also to overseas manufacturers in the USA or the Far East wanting to reach the whole of Europe using one integrated operation throughout.

Shipments can be collected at origin, flown on our own aircraft or sent by sea to Europe with the goods then kept under bonded FEMBAC licence (or CBAC licence, as it is now called) in our own warehouse. On arrival, we provide a full inward goods management system, updating the shipper's host system with any discrepancies.

The goods are then checked and stored in our warehouse, using FIFO (first-in-first-out) procedures, with perpetual inventory management and so on.

On receipt of individual customer orders, we pick and pack and carry out quality checks, including reconfiguration where this is required, and then prepare all the necessary Customs documentation for release from the bonded warehouse as well as the export documentation for the final transit.

The goods are then sent by either road or air throughout Europe, still within the overall integrated operation, with full tracking control and with POD's being updated on our worldwide system.

Finally, I would like to give you an example of two case studies from customers we currently handle on a pan-European basis. For reasons of confidentiality the company names are not mentioned, but the first is a main manufacturer of PCs who distributes throughout Europe to provide a next-day response to customer orders.

Under this operation, we cover every address within Europe and the service is especially tailored to the individual requirements of the customer, including a combined bar-coded label which is used by both ourselves and the customer. We also have special billing arrangements and management reports showing the effectiveness of on-time delivery.

The other example is that of a major automotive manufacturer with major plants at various locations throughout Europe. Spares are distributed from each of these plants to all countries in western Europe and some eastern European locations.

The delivery time is very specific with timed a.m. deliveries which must be effective to within one hour of the committed time.

In some countries, deliveries are to a central location, but in others deliveries are made direct to dealers on a nationwide basis.

A particular feature of this service is the provision of proactive notification to each final consignee prior to delivery should a shipment for any reason be delayed. In addition there is a full 100% monitoring of all shipments, giving the exact reason for non-delivery, for example, weather, customs, strikes and so on. And at the end of the month, this information is provided in the form of bar charts showing the exact performance against the targeted levels and providing a full analysis of all deliveries. This enables us to discuss on a regular basis ways of further improving reliability.

So what can we learn from all this about the future changes and structures of our parcel delivery systems in Europe? Slowly the borders are coming down and the previous Customs procedures are disappearing – initially between EC countries but, hopefully before long, throughout Europe. Most companies are beginning to regard Europe as single market and, slowly and painfully, the barriers to free competition are being dismantled. As in the USA, full logistics service providers are emerging with more companies contracting out their distribution. Door-to-door tracking systems are essential for success, with companies in Europe increasingly demanding effective measurement of performance. I believe reliable, time-controlled transport will continue to remain the key to success.

The traditional structure of parcel groupage and forwarding operations in Europe will change just as rapidly as they have in the USA and, with this change, I perceive service levels improving year-by-year. At the same time, we shall see rates coming down from current levels to far nearer costs of similar movements in the USA.

As I said earlier, I have never been involved in such a period of dramatic change. It's an exciting time to be in the industry and provided that free competition is allowed, with no unfair cross subsidisation at any level, then I believe the market has huge potential. Whether you are an integrator, or a multi-national company or an IT service provider, I believe there are great opportunities for us all.

16 Quick Response in Retail Distribution: An International Perspective

J. Fernie
Institute for Retail Studies
University of Sterling

1 INTRODUCTION

This chapter discusses the implementation of the concept of Quick Response to different retail markets of the world, notably the UK, USA, continental Europe and Japan. While much of the initial work on Quick Response focused upon the fashion sector of business, this chapter deals specifically with grocery markets where arguably Quick Response should be a part of corporate philosophy. It will be shown that the enabling technologies to implement Quick Response are in place but success at reducing inventory through the supply chain and in minimizing lead times varies not only from country to country but between companies in specific countries. The reasons for such variations include the nature of retailer – supplier relations, the degree of fragmentation or concentration of retail markets, the extent of retail branding and the distribution 'culture' evident in different parts of the world.

2 QUICK RESPONSE: THE ENABLING TECHNOLOGIES

Quick Response was a phrase coined by management executives in the USA to speed up the response time from a customer choosing a fashion item to replenishing it right throughout the supply chain. The fashion business is volatile and it was often difficult to realize the logistician's dream of getting the right product to the right place at the right time. Clearly any improvements in supply chain efficiencies would impact upon the bottom line and the trade press and other literature sources are replete with examples of cases of increased sales revenues and massive cost savings (see examples quoted in Walker, 1994a, b; Fox, 1991; Gill, 1991; Retail Perspectives, Quarter 1, 1992; Connections, 5, 1993). The grocery industry is different from the clothing industry; it should be practising quick response as a matter of its logistics policy. The majority of grocery items are relatively stable in terms of consumer demand and, apart from seasonal peaks, forecasting product flow through the supply chain is more of an exact science than in the fashion business. The concept of Quick Response is therefore given and attention has focused upon how much more efficiency can be squeezed out of the supply chain. The new Buzz phrases of the 1990s for the grocery sector are time compression (Galloway, 1991), efficient consumer response (Kurt Salmon, 1993) and fast flow replenishment systems (Anderson Consulting, 1992).

The enabling technologies to implement Quick Response initiatives have been available for some time but the adoption of the technologies has been slow – even among the innovators. At the EDI'93 Conference and Exhibition, Sir John Harvey-Jones chastised British Industry for its slow adoption of this particular technology despite the fact that British companies, including those from the retail sector, lead the rest of Europe in the implementation of EDI (Walker, 1994b). But the problem with Quick Response is that it must not be viewed as one technology but a series of technologies with uniform product codes being scanned at the checkout to yield meaningful data on customer choice, EDI linking partners through the supply chain in the transmission, receipt and payment of orders and the scanning of products at warehouses through the supply chain to track goods from factory to final customer.

In a survey of UK retailers, Bamfield (1993) has traced the adoption of EDI from the early 1980s to the present day. His research suggests that EDI uptake was slow at first then accelerated beyond expectations by the early 1990s. His reasons for such a change in perspective in such a short period of time were due to:

- EDI innovative networks;
- cost and performance improvements; and
- management learning.

Bamfield (1993) maintains that retailers began to recognize the benefits of EDI via their interest in developing EPOS systems. The acceptance of common standards in bar coding therefore preceded an interest in trials of EDI information systems. At the same time IT departments, invariably trained on centralized batch systems, had undergone considerable change and training through the implementation of EPOS systems. In terms of adoption, the Quick Response technologies were beginning to appeal to the key decision makers of retail companies as the risks of adoption were minimized and benefits were likely to be realized. As costs of transmitting data continue to fall and greater standardization of third party networks occur, more and more smaller companies can achieve the benefits of Quick Response.

Although Machell (1993) claims that US retailers are more advanced than their European counterparts in their adoption of Quick Response technologies, research indicates that similar barriers to adoption exist in the US to those in the UK (see Figure 16.1). This figure is derived from the Horizon Scan Project which seeks to identify the impact of EDI and other QR technologies on product identification systems in the retail industry. Similar reasons for slow adoption of these technologies are given to those by Bamfield (1993) in the UK, namely a lack of industry standards with many larger retailer organizations using different value added net-

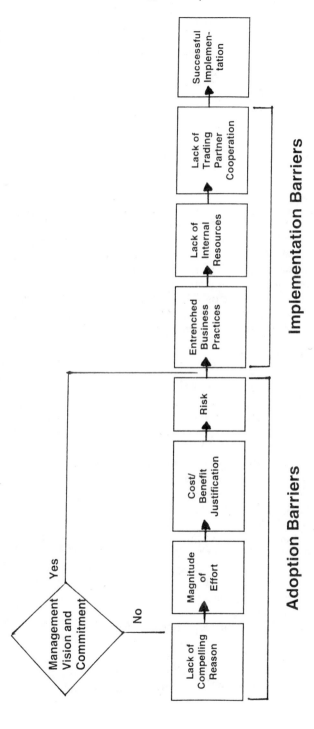

FIGURE 16.1. Barriers to technology adoption.

Source: Uniform Code Council, Horizon Scan Project

works (VANS) to carry their EDI transactions, the cost involved for small vendors and a degree of management inertia to new technology (McCusker, 1993; Davis, 1989). While most of these barriers should be overcome, research in both the UK and US would appear to suggest that the main obstacle to QR may well be in the area of sharing EPOS-derived demand data for forecasting production planning because of confidentiality and other constraints (Bamfield, 1993; Whiteoak, 1993; Mercer Consulting, 1993).

3   QUICK RESPONSE: EMPIRICAL EVIDENCE

3.1   The USA

The most thorough review of the grocery retailing sector in its adoption of QR technologies has emanated from the USA, most notably two reports by consultants Kurt Salmon and Mercer Management Consulting. The Kurt Salmon report on Efficient Consumer Response has been championed as the blueprint for successful supply chain management in the US to the extent that its proposals have been mooted for adoption in Europe (Logistics Europe, October, 1993). While this may be the case for continental Europe, it will be shown that grocery retailers in the UK have already implemented many of the recommendations contained in the report.

While the scale of the savings envisaged are vast – $10 billion in the dry grocery chain – it is easy to appreciate how such savings can be achieved in view of the inefficiencies of the existing supply chain. Kurt Salmon was commissioned to conduct this analysis because of the US grocery industry's declining profitability aggravated by increased competition from the price competitive mass merchants and warehouse clubs. Figure 16.2 shows the current situation with regard to lead times in the dry grocery supply chain; it takes 104 days, on average, for such a product to pass from the supplier's packing line to the consumer at the checkout. If perishable products are included the lead time is reduced to 75–80 days. This equates with a stock turn of three and a half times for dry grocery and around five times for dry/perishables. The efficiencies alluded to in the report would increase dry grocery turnover to six times.

One of the main reasons for such a large volume of stock in the supply chain pipeline is the fragmentation of the chain especially in the replenishment process. Product currently flows through three independent information and product flows:

- from store to consumer;
- from distributor warehouse to store; and
- supplier to distributor warehouse.

Stock is replenished by being 'pulled' through the supply chain by store replenishment orders but inventory tends to be 'pushed' into the warehouse by forward buying and trade deals. Forward buying for trade deals was mainly used

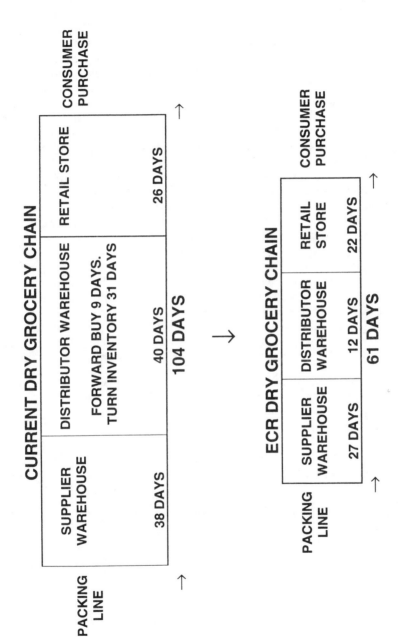

FIGURE 16.2. Comparison of average throughput time of dry grocery chain before and after ECR implementation. Source: Kurt salmon (1993).

in the inflationary 1970s because suppliers, fearing price controls, maintained high prices but offered discounts to distributors who in turn were able to increase gross margins through shrewd buying. Unfortunately forward buying, which was used as a competitive weapon in the 1970s, became the norm for the retail sector in the 1980s and 1990s. In terms of grocery suppliers' promotional spend between 1981 and 1991 advertising has declined from 43% of overall spend to 25% whilst trade promotions has increased from 34% to 50% (consumer promotions has remained stable at around 25%). What this means is that the original *raison d'être* for trade promotions has less validity in the 1990s but the costs of such a strategy continues to rise. Kurt Salmon (1993) argues that the increasing complexity of deals involves considerable administrative costs with some retailers having 7–8000 deals on file at any one time and buyers/sales representatives spending 10–15% of their time resolving pricing discrepancies. Trade promotions also result in high variability and unpredictability in demand. These peaks and troughs in the volume of product passing through the supply chain lead to excessive inventories and inefficient utilization of warehouse and transport capacity.

Further administrative costs are encountered because of consumer promotions, primarily coupon redemption. Over 280 billion coupons are distributed annually in the USA although only 2.6% of this total is redeemed. The processing and validation of such promotions are under review in order to EPOS available coupons for distributor reimbursement.

At the store level more efficient store assortments are proposed in addition to improving backroom space efficiency. For example, Table 16.1 highlights the savings that can be achieved within the store. Much of the stock held in store is a result of forward buying initiatives discussed above and more backroom space

TABLE 16.1. Sales benefit or efficient space management.
Source: Kurt Salmon (1993)

Space Description	Current Size (sq feet)	Sales per sq. ft/week	Sales/week	Efficient Space Management Improvement	Freed Space (sq feet)	Increased Sales/week
Selling Space	27,500 (75%)	$8.75	$241,000	Same Volume in 10% Less Space	2,750	$24,062
Non-Selling	9,100 (25%)	$0	$0	Reduced Need to 20% of total store	1,780	$15,575
TOTAL	36,600	*	$241,000	*	4,530	$39,637 or 16% of Sales

*Assumes average sales per square foot

will be released as selling space as moves to continuous replenishment gather pace. Overall the proposed efficient consumer response strategies proposed by Kurt Salmon embrace all elements of the supply chain from utilizing category management at store level to improving replenishment cycles and enhancing promotion and new product introductions. Clearly the benefits from implementing these recommendations will lead to greater customer satisfaction and improved profitability for suppliers, wholesalers and retailers if collaboration between these parties can be achieved to improve efficiency throughout the supply chain.

3.2 The UK

Much of what is contained in these reports on the US scene will not appear to be particularly innovative to the UK grocery retail sector. With the progressive adoption of the enabling technologies, the major multiple retail groups have made significant improvements in cutting inventories and reducing lead times.

All the major grocery retailers have over 90% of their products centralized through a network of Regional Distribution Centres (RDCs). The upgrading of depot networks to push more product through newly constructed RDCs was the first step by UK grocery retailers to embrace the Quick Response concept. This enabled depots to provide stores with daily deliveries of product and also progressively reduced lead times below 24 hours for much of a company's product range. Furthermore, the growth of purpose-built multi-temperature 'composite' distribution centres and compartmentalized trailor units allowed the 'majors' to streamline deliveries whereby large RDCs would deliver to large superstores in large tractor/trailers (see Table 16.2).

TABLE 16.2. Profile of store and depot characteristics of UK grocery retailing 1995.

Operator	No. of Stores	Average Selling Space of Store (000, sq ft)	No. of Depots	Size Range (000, sq ft)	No. of Composites
Argyll Group	551	16.5			
Safeway	373	21.5	13	30–510	4
Presto	178	6.1			
Asda	203	40.3	10	63–465	6
CRS	436	5.8	11	17–185	0
J Sainsbury	346	25.9	19	34–427	7
Tesco	432	25.6	17	130–450	7

Further efficiencies in supply chain management are being achieved by the multiple grocers. The notion of 'nil' stock for fast moving, short shelf life products (under 21 days) is being implemented by the industry leaders as these products are cross-docked in RDCs for onward delivery to stores. In essence, stock only exists in store or in transit for these products. The implications for warehouse and transport management of these changes are profound. The internal configuration of RDCs are changing as 'reserve' slots on upper racking are giving way to more 'picking' slots leading to semi-automation and redesigning of warehouse space. In terms of transport management, the demand for more frequent deliveries of smaller quantities from suppliers is causing manufacturers to review existing transport arrangements in that common user transport is making a comeback in the UK.

The UK 'majors' are also tying more and more of their suppliers into EDI networks with Tesco having 1200 of their 2500 suppliers linked to facilitate the transmission of orders, invoices and payments. Similarly Sainsbury are linked to 700 of their suppliers, Asda with 595 (EAN, 1993). At the front end of the supply chain, sales based ordering is being rolled out by the multiples to use EPoS data more effectively than hitherto in accurately forecasting individual store demand by SKU (Nixon, 1993; Hogarth-Scott and Parkinson, 1993; Sanghavi, 1993). The eventual success of these programmes, however, depends heavily on the degree of cooperation between retailer and supplier.

Despite much of the rhetoric about partnershipping the practice of working closer together to share risks as well as the benefits still has a long way to go. Whiteoak (1993) has shown that unless the majority of retailers share information, production schedules will still be planned with a high margin of error as lead times shrink. Clearly much remains to be done in breaking down attitudes and prejudices from the past to a more open approach of securing common benefits in supply chain coordination. Williams (1993) of Safeway recognizes these difficulties and believes that as both retailers and suppliers experience diminishing returns, they will be forced to work more closely together.

3.3   The rest of the EU

The extent to which the UK companies are further down the experience curve in logistics management was confirmed by a major survey undertaken by GEA Consulentia Associata in 1994. Based on the Kurt Salmon methodology to ensure compatibility of results, GEA (1994) show that efficient consumer response programmes will yield less benefits in Europe than in the US, for example savings of 2.3 to 3.4 percent of sales was estimated compared with 10.8 percent quoted by Kurt Salmon. Furthermore, Table 16.3 shows that companies in all four European countries hold much less stock than their counterparts in the US. Even then, companies hold minimal stock in store and at RDCs compared with the other three countries where the forward buying of promotional stock is still widely practiced (Bedeman, 1989; Pellew, 1993). A recent survey by Paché (1995) confirms that this practice continues to be used by major French retailers although there are signs of a move to implement efficient consumer response programmes.

TABLE 16.3. Total channel stock ambient products (days).
Source: Walker (1994)

Country	TOTAL CHANNEL STOCK (Days)			
	Italy	France	UK	Germany
Supplier	12	14	11	14
RDC	19	20	11	22
Store	9	9	7	11
TOTAL	40	43	29	47

The concept of composite and dedicated distribution is rare on mainland Europe whereas it has become a major feature of UK grocery distribution. The development of composites in the UK resulted from the close working relationships that distribution service companies nurtured with their retail clients. Over the last 10 to 15 years these companies moved from being road hauliers to providers of logistics services and dedicated contracts were provided by retailers as they contracted out more and more of their distribution. In the rest of Europe retailers have tended to build product specific warehouses, run 'in-house'. Transport is treated more as a commodity bought on a 'spot-price' basis. Cooper *et al.* (1992) have questioned the efficiency of dealing with such a range of transport companies in view of the administrative costs involved. This approach to the purchasing of logistics services is very similar to that in the USA (Penman, 1991).

3.4 Japan

The Japanese system of retail distribution is relatively unique compared with either the US or European examples discussed above. The Japanese consumer's passion for new products means that retailers stock a proliferation of lines with little back-up inventory in store. This places considerable strain on the logistical network especially as marketing channels are highly fragmented with a multi-layered system of manufacturers, wholesalers and retailers. Unlike in Europe or North America, marketing is more about developing long term relationships with channel partners than selling to consumers and this 'Eigyou' system results in an efficient and costly supply chain. Ohboro, Parsons and Riesenbeck (1992) claim that Japanese consumers pay for this inefficiency through prices 40% greater than those in the USA and 35% greater than those in Germany. There are indications, however, that retailers and particularly discounters will begin to deal directly with manufacturers in an attempt to lower prices to the customer (The Grocer, 16 January 1993).

4   EXPLAINING THE DIFFERENCES

It is clear that distribution networks vary considerably within and between countries. The main factors which tend to explain these variations are:

- the extent of retail power;
- the penetration of store brands in the market;
- the degree of supply chain control;
- types of trading format;
- geographical spread of stores;
- relative logistics cost; and
- relative sophistication of the distribution industry.

4.1   Retail power

There has been a significant shift in the balance of power between manufacturer and retailer during the last 20–30 years as retailers increasingly take over res-ponsibility for aspects of the value-added chain, namely product development, branding, packaging and marketing. These changes in supplier-retailer relation-ships vary in time and space according to the extent of retail concentration and the degree of fragmentation of suppliers' markets. Ohboro, Parsons and Riesenbeck (1992) maintain that this power struggle is finely poised in the US where a combination of brand pull and innovative product development by manufacturers ensure that suppliers' products reach supermarket shelves. Furthermore, the US grocery retail market is largely regional in character and national grocery manu-facturers can still wield power in the market place. By contrast in Europe and Japan the balance of power is uneven. In Japan the 'Eigyou' encourages a manu-facturing driven system which has been reinforced through a strongly regulated retail sector which has led to a large number of small stores. Thus the Japanese distribution channels are fragmented and inefficient, for example Japan has 620,000 food stores serving a population half of that of the US which has 145,000 outlets.

   The situation in Japan is markedly in contrast to that of much of northern Europe where grocery retailers have grown rapidly from regional to national and international concerns to the extent that a high degree of concentration exists in these retail markets. While the actual measurement of market shares is fraught with statistical inconsistencies, it is generally agreed that companies in the UK, Germany and the Benelux countries have the highest concentration levels and the southern European nations, Greece, Italy and Portugal, the more fragmented retail markets (Wileman, 1992; Fernie, 1994).

4.2   Retail branding

Commensurate with the growth of these powerful retailers has been the develop-ment of distributor labels. Much debate has centred on the type of brand (Davies,

1992; Pellegrini, 1993) or the strategies open to suppliers (Wileman, 1992; Glemet and Mira, 1993). The intensifying price war in the UK has re-focused retailers' strategies partly back to more value for money tertiary brands than hitherto, but the UK scene is still dominated by value-added store brands compare with the situation in the rest of mainland Europe where cheaper alternatives to manufacturers brands are the norm (Samways, 1995). Regardless of the type of distributor label, the manufacturer is supplying brands exclusively manufactured for the retail trade. Wileman (1992) maintains that concentration and therefore power is a necessary pre-requisite before developments such as own label.

4.3   Supply chain control

Firms in countries with greater concentration in their grocery markets and a high level of own label penetration tend to assume responsibility for distribution support to their stores. In the UK the transition from a supplier driven system to one of retail control is complete compared with other parts of Europe, US and Japan. In Europe retail control is more evident in the Benelux countries, followed by France, Germany and the Scandinavian countries with the southern European countries continuing to be dominated by a supplier-led distribution system (Wileman, 1992; Jourdan and Irving, 1992; Cullis, 1992). Wileman (1992) states that it is still possible to find sales and merchandising forces of 3000 people providing direct coverage of 80–100,000 retail outlets in Italy and Spain.

4.4   Trading formats

It was shown that the evolution of distribution networks in the UK was partly driven by the rise in prominence of the superstore as the dominant trading format. Although the character of the grocery market is changing with the increased competition from discounters, existing networks run by the 'majors' have been fashioned through the need to supply large stores from large RDCs, often composite in nature. In the rest of Europe, trading formats vary considerably from suprettes, discounters to hypermarkets and the major companies are invariably well represented in all these markets. For example, Tengelmann has 29 warehouses supplying its 3501 stores trading under different formats and fascias. Most grocery stores in mainland Europe are supplied by product category warehouses similar to the smaller supermarket operators in the UK. Albert Heijn, the major grocery chain in the Netherlands, opened a centralized, composite facility in 1993 but their perception of a composite is to house a variety of product category warehouses on the one site. This is similar in approach to that in the US; indeed, Safeway's Aylesford site in Kent is a legacy from the US operation and is different from the Argyll Group's approach to composite distribution.

4.5   Geographical spread of stores

The size and spread of stores will largely determine the form of logistical support to the retail operation. For example, the limited line discounters pose few problems to warehouse managers but the transport operation is more complex in view of the large number of stores served. By contrast, the retail operators with wide product ranges require strategies for slow movers and configuration within the warehouse to meet Quick Response schedules. In order to increase efficiency in the transport operation, delivery schedules are planned to either maximize internal or external backloading.

At a national level the actual physical distances that have to be covered varies from countries such as the UK and the Benelux countries to that in the US and to a lesser extent France and Spain. Centralization and the development of RDCs were more appropriate to the geography of the highly urbanized environments of much of northern Europe. Hence, Cooper *et al.* (1992) argue that the RDC concept has limited application to countries such as France and Spain where the hypermarket is a major trading format. Here, the hypermarket has a stand alone status because of the distances involved to add another link in the supply chain.

4.6   Logistics costs

The make up of the components of logistics costs is complex and dependent upon a range of factors relatively unique to individual countries. In a simplified way these costs are influenced by the classical factors of land, labour and capital. It is not surprising that the countries with highest urban densities have moved further down the route to fast flow replenishment systems. High land and property costs have acted as an incentive for Japanese, UK and Benelux operators to maximize sales space and minimize the carrying of stock in store. Indeed, in line with Porter's (1990) reasoning that competitive advantage often comes through having to innovate, this has been the case in logistical innovation. JIT has its origins in Japan, warehouse automation has been mainly developed in the Scandinavian countries because of high labour costs and, arguably, UK grocery retailers have been innovators in Quick Response because of relatively high interest rates compared with other countries in the late 1970s and 1980s.

4.7   The role of the contractor

Cooper *et al.* (1992) and Penman (1991) have shown that the UK is fairly unique in that professional contractors provide a dedicated distribution service to retailers, that is the management of RDCs and transport to stores, compared with a more fragmented support function found in the rest of Europe and the USA. In mainland Europe and the US most warehousing is own account while the transportation function is contracted out, often to local hauliers. Trucking in the US is predominantly carried out by owner drivers and deregulation of transport markets has led to competitive pricing for freight services. Similarly the progressive deregulation of EU markets should foster greater competition although the break-down of nationally protected markets has been slow and Europe continues to be dominated by general haulage rather than more specialist services.

It can also be argued that the millions of pounds invested in composite distri-bution requires a more sophisticated distribution support role from contractors than the mere running of low specification 'sheds'; indeed, third party operators in the UK have been in the vanguard of technological innovation in the retail sector. This and other reasons, mainly of a financial nature, have led to a greater degree of contracting out in the UK compared with other countries to the extent that leading UK distribution companies such as Exel Logistics and Christian Salvesen have sought to exploit opportunities in mainland Europe and the US, often with British companies expanding abroad.

5   SUMMARY

This chapter has attempted to assess the impact of Quick Response in various grocery markets of the world. It was shown that the enabling technologies are available but even the innovators have been slow to adopt them until standard-ization of both hardware and software was achieved. Empirical evidence indicates the differences in approach in different countries. Broadly, companies in the US and many of the countries of mainland Europe can achieve greater efficiencies throughout the supply chain by implementing Quick Response techniques. The supply chain in these markets is fragmented, products are pushed through distri-bution channels and forward buying and trade promotions continue to be a feature of supplier-distributor relations. By contrast the UK market already embraces the concept of Quick Response and the major companies are now trying to achieve even greater efficiencies through partnership arrangements with their suppliers. Explanations for such variations in international markets were discussed. It was shown that the greater the concentration of retail markets, the higher the level of retail branding with retailers pulling stock through the supply chain. As retailers control the supply chain they centralize stock in RDCs, often managed by third party contractors. At a micro level, however, distribution support to stores is largely influenced by the type and number of stores which have to be supplied. This determines the type and number of warehouses required within the logistical cost framework for individual countries.

REFERENCES

Anderson Consulting (1992) Grocery Distribution in the 90s; strategies for fast flow replenishment, Coca Cola Retailing Research Group.

Bamfield, J. (1993) Technological management learning: the adoption of EDI by retailers, paper presented at the 7th International Conference on Research in the Distributive Trades, University of Stirling.

Bedeman, M. (1990) Logistics options in Europe, paper presented at a British Retailer's Association Conference, *Logistics – the European Impact*, London.

Cooper, J. Browne, M. and Peters, M. (1992) *European Logistics: Markets, Management and Strategy*, Blackwell, Oxford.

Cullis, R. (1992) European Distribution Services, IGD, Watford.

Davis, L. (1989) Retailers go shopping for EDI, *Datamation*, 35, March, pp. 34-36.

Davies, G. (1992) The two ways that retailers can be brands, *International Journal of Retail & Distribution Management*, 18(2), pp. 24-34.

EAN International (1993) Electronic Data Interchange in the EAN Community, EAN, Brussels.

Fernie, J. (1994) Retail Logistics, in: (Ed. J. Cooper) *Logistics and Distribution Planning*, Kogan Page.

Fox, B. (1991) KG Retail sets the pace, *Chain Store Executive Age*, 67, January, pp. 94-97.

Galloway, J. (1991) Asda's central distribution system over the last three years, Exel Logistics Distribution Forum, 26 July.

GEA (1994) Supplier – Retailer Collaboration in Supply Chain Management, Coca-Cola Retailing Research Group, Europe.

Gill, P. (1991) QR keeps jeans moving, *Stores*, 73, February, pp. 23-24.

Glemet, F. and Mira, R. (1993) The brand leader's dilemma, *The McKinsey Quarterly*, 2, pp. 3-5.

Hogarth-Scott, S. and Parkinson, S. T. (1993) Retailer-supplier relationships in the food channel – a supplier perspective, *International Journal of Retail & Distribution Management*, 21(8), pp. 12-19.

Jourdan, P. and Irving, R. (1992) Food distribution in the UK, continental Europe and the US, paper presented at: *Integrating the supply chain*, Rotterdam.

Salmon, K. (1993) Efficient Consumer Response: enhancing consumer value in the supply chain, Kurt Salmon, Washington.

McCusker, T. (1993) How to get from 80 to 100% in EDI, *Datamation*, 39(3), pp. 45-48.

Machel, P. G. (1993) Supply chain strategy in a multinational environment, paper presented at an IGD conference on *Improving supply chain effectiveness in the grocery trade*, London.

Mercer Management Consulting (1993) New Ways to Take Costs out of the Retail Food Pipeline, Coca Cola Research Group, Atlanta.

Nixon, I. P. (1993) Innovative IT applications in supply chain management, paper presented at an IGD conference, op. cit.

Ohboro, T., Parsons, A. and Riesenbeck, H. (1992) Alternative routes to global marketing, *The McKinsey Quarterly*, 3, pp. 52-74.

Paché, G. (1995) Speculative Inventories in the Food Retailing Industry: a Comment on French Practices, *International Journal of Retail & Distribution Management*, 23(12), pp. 36-42.

Pellegrini, L. (1993) Retailer brands: a state of the art review, paper presented at the 7th International Conference, Stirling, op. cit.

Pellew, M. (1993) An holistic approach, *Logistics Europe*, 1(5), pp. 37-39.

Penman, I. (1991) Logistics – fragmented or integrated, *Focus*, 10, (9), pp. 21-24.

Porter, M. E. (1990) *The Competitive Advantage of Nations*, MacMillan.

Samways, A. (1995) Private Label in Europe, FT Management Report, Pearson.

Sanghavi, N. (1993) The buyer/seller relationship in UK retailing, paper presented at the 7th International Conference, Stirling, op.cit.

Walker, M. (1994a) Quick Response: the road to lean logistics, chapter in Cooper, op. cit.

Walker, M. (1994b) Supplier – retailer collaboration in European grocery distribution, paper promoted at an IGD Conference *Profitable Collaboration in Supply Chain Management*, IGD, London.

Whiteoak, P. (1993) The realities of quick response in the grocery sector – a supplier viewpoint, *International Journal of Retail & Distribution Management*, 21(8), pp. 3-11.

Wileman, A. (1992) The shift in balance of power between retailers and manufacturers, paper delivered at Management Centre Europe, Brussels, June.

Williams, R. (1993) The realities of quick response in the grocery sector: retail perspective, paper presented at an IGD conference, op. cit.

Printing: Weihert-Druck GmbH, Darmstadt
Binding: Buchbinderei Schäffer, Grünstadt